SpringerBriefs in Astronomy

SpringerBriefs in Astronomy are a series of slim high-quality publications encompassing the entire spectrum of Astronomy, Astrophysics, Astrophysical Cosmology, Planetary and Space Science, Astrobiology as well as History of Astronomy. Manuscripts for SpringerBriefs in Astronomy will be evaluated by Springer and by members of the Editorial Board. Proposals and other communication should be sent to your Publishing Editors at Springer.

Featuring compact volumes of 50 to 125 pages (approximately 20,000–45,000 words), Briefs are shorter than a conventional book but longer than a journal article. Thus Briefs serve as timely, concise tools for students, researchers, and professionals.

Typical texts for publication might include:

- A snapshot review of the current state of a hot or emerging field
- A concise introduction to core concepts that students must understand in order to make independent contributions
- An extended research report giving more details and discussion than is possible in a conventional journal article
- A manual describing underlying principles and best practices for an experimental technique
- An essay exploring new ideas within astronomy and related areas, or broader topics such as science and society

Briefs allow authors to present their ideas and readers to absorb them with minimal time investment.

Briefs will be published as part of Springer's eBook collection, with millions of readers worldwide. In addition, they will be available, just like other books, for individual print and electronic purchase.

Briefs are characterized by fast, global electronic dissemination, straightforward publishing agreements, easy-to-use manuscript preparation and formatting guidelines, and expedited production schedules. We aim for publication 8–12 weeks after acceptance.

More information about this series at http://www.springer.com/series/10090

Jeremy Wood

The Dynamics of Small Solar System Bodies

 Springer

Jeremy Wood
Hazard Community and Technical College
Hazard, KY, USA

ISSN 2191-9100   ISSN 2191-9119   (electronic)
SpringerBriefs in Astronomy
ISBN 978-3-030-28108-3   ISBN 978-3-030-28109-0   (eBook)
https://doi.org/10.1007/978-3-030-28109-0

This Springer imprint is published by the registered company Springer Nature Switzerland AG.
The registered company address is: Gewerbestrasse 11, 6330 Cham, Switzerland

Acknowledgments

Tremendous gratitude goes to Associate Professor Stephen C. Marsden, Professor Jonathan Horner, and Dr. Tobias C. Hinse for their feedback.

Contents

List of Figures

List of Tables

Chapter 1
Introduction

1.1 The Beginning

From antiquity, human beings have been looking upward at the heavens. The first humans saw the Sun, the Moon and stars that formed shapes in the sky. Every day and night these heavenly objects were seen to constantly move across the sky in an east-to-west motion relative to the horizon.

As this motion occurred, most stars maintained their position relative to other stars, but some stars were seen to wander—alternating between eastward and westward motion relative to other stars. These wanderers were not stars at all, but planets.

These planets would have been indistinguishable from stars were it not for their unusual motion. Though they would become difficult or impossible to see when near the Sun in the sky, they would always move away from the Sun again and return to better visibility.

But stars and planets were not the only objects our ancient ancestors observed. Occasionally, these ancient people noticed strange, fuzzy objects that appeared in the sky in places where no object had been seen before. They noticed that these surprise guests had tails and were larger than any star. Over time, these objects would fade from view and disappear—even when far from the Sun in the sky. These were the first human sightings of comets—one type of small body in the solar system.

In a sense, the science of small solar system bodies began with the very first comet sighting. It would be millennia, however, before the exact nature of comets would be understood, and along the way other types of small bodies would be found.

© The Author(s), under exclusive license to Springer Nature Switzerland AG 2019
J. Wood, *The Dynamics of Small Solar System Bodies*, SpringerBriefs
in Astronomy, https://doi.org/10.1007/978-3-030-28109-0_1

1.2   What Is a Small Solar System Body?

Early humans never fully understood the exact nature of these strange visitors in the night sky. From those humble beginnings, we now know that small bodies are members of our own solar system and can be found anywhere from inside Earth's orbit to outside the orbit of Neptune, the most distant planet from the Sun.

Two major types of small solar system bodies are **comets** and **asteroids**. Traditionally, the difference between a comet and an asteroid has been that a comet is composed of significant amounts of ices mixed with rock, and an asteroid is composed primarily of rock with little to no ice.

Today, we know the situation is more complicated. These various bodies differ greatly among each other in size, composition and orbital characteristics. It is convenient to group and name a subset of small bodies that share common characteristics so that they may be more easily referred to.

Dozens of such populations exist and are classified based on physical properties or other characteristics. For example, a group of rocky Earth-crossing objects that orbit the Sun interior to Earth's orbit are collectively called the Atens.

To differentiate such populations from other bodies in the solar system, a small solar system body (or SSSB) must have a definition. It is tempting to state that all bodies with sizes smaller than planets are SSSBs. But defining a SSSB based on size alone is problematic if planetary **satellites** are included in this definition, because some satellites are actually larger than our smallest planet. For example, Jupiter's satellite (or moon) Ganymede and Saturn's satellite Titan are both larger than the planet Mercury.

Also, since satellites orbit planets and not the Sun directly, they move along with their parent planet in its journey around the Sun. Other non-satellite small bodies orbit the Sun and are not attached to one particular planet. Generally speaking, this makes the orbital evolution of these bodies much different than that of satellites and suggests that satellites should be in a different category.

Other factors that affect the orbital evolution of small bodies include collisions and non-gravitational forces, such as forces due to solar wind pressure. Very small objects orbiting the Sun, such as rocks the size of your hand, or dust, are much more susceptible to these non-gravitational forces. For example, if a rock collides with a body many orders of magnitude larger in size, it can be swallowed whole and become part of the larger body. This is one reason why these very small objects have a different classification of their own.

If the size (and along with it, mass) of an object orbiting the Sun is low enough, it is classified as a **meteoroid** rather than a SSSB. Meteoroids include dust and rocks formed by collisions between other larger small bodies. No cutoff size exists below which a SSSB is considered to be meteoroid. Most are less than 10 m across.

When a meteoroid enters Earth's atmosphere, it is called a **meteor**, and if it hits the surface, a **meteorite**. Excluding satellites and meteoroids would leave all other objects smaller than planets orbiting the Sun classified as SSSBs. However, in 2006 the International Astronomical Union defined a brand new type of body called a **dwarf planet**.

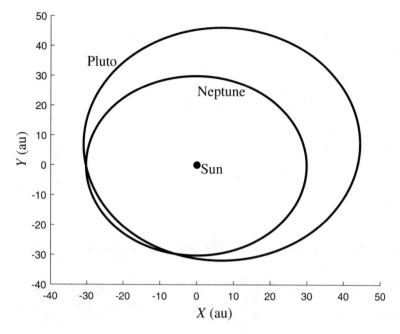

Fig. 1.1 Shown are the orbits of Pluto and Neptune around the Sun in two dimensions. The third dimension points out of the plane. Note how the orbit of Neptune crosses the orbit of Pluto. The Sun is located at the origin

A dwarf planet is massive enough to be spherical, is not a satellite, and orbits the Sun, but differs from a planet in one respect: it has not cleared the neighborhood of its orbit. To have its neighborhood cleared, a body must have a mass that is at least 100 times larger than the combined mass of all the objects that regularly cross its orbit. Pluto was demoted from a planet to a dwarf planet in 2006 because the more massive planet Neptune crosses its orbit. This is shown in Fig. 1.1.

The masses, mean radii and densities of all eight planets are shown in Table 1.1. For comparison, the masses and mean radii of a few of the largest small solar system bodies are shown in Table 1.2.

Finally, we can define a **small solar system body** as a natural body that orbits the Sun, is not a planet, is not a satellite, is too massive to be a meteoroid and is not a dwarf planet.

Example Masses and radii of celestial objects can be given in units relative to Earth's values to give a sense of relative mass and size. The dwarf planet Ceres has a mean radius of 473 km and a mass of 9.393×10^{20} kg. Given that Earth's mass and mean radius are 5.98×10^{24} kg and 6378 km, respectively, express the mass and mean radius of Ceres relative to Earth's corresponding mass and mean radius.

Table 1.1 The masses, mean radii and densities of the planets

Planet	Mass (Earth = 1)	Mean radius (Earth = 1)	Density kg m^{-3}
Mercury	0.0552	0.383	5427
Venus	0.814	0.949	5243
Earth	1	1	5514
Mars	0.107	0.532	3933.5
Jupiter	317.8	11.2	1326
Saturn	95	9.45	687
Uranus	14.5	4.01	1270
Neptune	17.1	3.88	1638

Masses and radii are given in units relative to those of Earth, which are 5.98×10^{24} kg and 6378 km respectively

Table 1.2 The masses and mean radii of a few of the largest small solar system bodies given in units relative to those of Earth which are 5.98×10^{24} kg and 6378 km respectively

Body	Mass (Earth = 1)	Mean radius (Earth = 1)
Pluto	0.00220	0.1868
Chiron	≤ 0.00000051	≤ 0.029
Chariklo	≤ 0.00000167	≤ 0.0215
Ceres	0.000157	0.0742

Solution

Relative Mass = $\frac{9.393 \times 10^{20}}{5.98 \times 10^{24}}$ = 0.000157

Relative Radius = $\frac{473}{6378}$ = 0.0742

Exercises

1. How can an observer on Earth tell the difference between a star and a comet viewed in the night sky?
2. How should a rock the size of your fist that orbits the Sun be best classified?
3. What change could be made to the orbit of Pluto to change it from a dwarf planet into a planet?
4. A rocky body shaped like a potato named Eros orbits the Sun between Mars and Jupiter with an unknown multitude of objects that regularly cross its orbit. Construct an argument against or in favor of Eros being classified as a dwarf planet.

5. Masses and radii of celestial objects can be given in units relative to Earth's values to give a sense of relative mass and size. The dwarf planet Haumea, which orbits outside the orbit of Neptune, has a mass of 4.01×10^{21} kg and a radius of 498 km. Express these values relative to Earth's corresponding mass and mean radius.

6. A student states that a small body of the solar system can be defined as any body that orbits the Sun and has a smaller size than any planet. Construct an argument against this statement, pointing out any flaws in reasoning.

Chapter 2
The Science of Small Solar System Bodies

2.1 Why Study Small Solar System Bodies?

On February 15, 2013, the city of Chelyabinsk, Russia was struck by a small rocky body about 20 m in diameter. The body exploded at a height of about 30 km above the ground. The resulting shockwave broke windows in over 7000 buildings and injured more than 1500 people. Some of the body survived the trip through Earth's atmosphere and made it to the surface.

Incredibly, the object was not detected before its impact with Earth's atmosphere. Other notable impacts occurred in Tunguska, Russia in 1908 and in the Yucatan 65 million years ago. The latter impact is believed to have caused the extinction of the dinosaurs and other species.

Given the destructive nature of such impacts, it behooves humanity to determine the threat such small bodies pose to Earth[1] (Hahn and Bailey 1990; Napier et al. 2015). Or as astrophysicist Neil Degrasse Tyson once said, "...if you're only looking down, one day the asteroid's coming ... at some point you got to look up."

Small bodies that may pose a collisional threat to Earth are given their own classifications. SSSBs in orbits that bring them into Earth's neighborhood are known as **Near Earth Objects**, or **NEOs**. The term **Near Earth Asteroids** or **NEAs** is also used.[2] Astronomers are dedicated to discovering and determining the orbits of bodies that may cause a threat to Earth's existence. Organizations at the forefront of this work include NASA and Pan-STARRS.

The United States government has directed NASA to find, track and catalogue at least 90% of the estimated population of NEOs that are equal to or greater than 140 m in size by 2020.[3] To date, more than 15,000 of these objects have been found.

---

[1] https://www.nasa.gov/mission_pages/asteroids/overview/index.html (accessed Dec. 28, 2017).

[2] https://cneos.jpl.nasa.gov/about/neo_groups.html.

[3] https://www.nasa.gov/planetarydefense/overview.

J. Wood, *The Dynamics of Small Solar System Bodies*, SpringerBriefs in Astronomy, https://doi.org/10.1007/978-3-030-28109-0_2

A subclass of NEOs called **Potentially Hazardous Objects** or **PHOs** (and **Potentially Hazardous Asteroids** or **PHAs**) are NEOs larger than 140 m whose orbits come within 0.05 au to Earth's orbit. Given two orbits and one point in each orbit. The minimum orbit intersection distance or **MOID** is the minimum distance between any two such points. Thus, for an object to be classified as a PHO, it does not necessarily need to approach Earth to within 0.05 au. Rather, the MOID of its orbit needs to be within 0.05 au of Earth's orbit.

Example The asteroid Apophis will approach Earth within a distance of 0.0002644 au in the year 2029. Should Apophis be classified as a potentially hazardous asteroid (or PHA)? Does this mean that the MOID of the orbit of Apophis must exactly equal 0.0002644 au?

Solution Since the orbit of Apophis will bring it closer to Earth than 0.05 au, it should be classified as a potentially hazardous asteroid. The MOID does not have to exactly equal 0.0002644 au and can be smaller than this value. The MOID has to do with the distance between orbits, not the distance between objects in those orbits.

But as Napier et al. (2015) have shown, a small body does not have to directly collide with Earth to be a threat. Gravitational forces from planets can perturb the orbits of small bodies, sending them outward or inward towards the inner solar system.

If a large (>100 km), icy body was to be perturbed into the inner solar system, it could spread large amounts of dangerous debris that could either collide with Earth or block sunlight (Hahn and Bailey 1990; Napier et al. 2015).

Ironically, the same types of small-body impacts that can destroy life can also bring life. Our solar system started out as a giant cloud of gas and dust known as the **solar nebula**, and within this cloud the Sun and planets formed.

Collisions were an essential part of the planet formation process. During their formation, planets like Earth increased in size and mass by accreting smaller objects and debris. In an analogous way, a snowball grows into a snowman by rolling in the snow. This process is called **accretion**.

During the early days of the formation of our solar system, the accretion process left Earth too hot for water to remain on its surface as a liquid. Yet, water abounds on Earth today, making up over 70% of its surface. One theory that explains the presence of water on Earth today is that after Earth cooled, comets collided with Earth over eons of time, bringing life-giving water and even organic compounds with them.

Today, it is well known that comets can contain substantial amounts of water in the form of water ice, as well as other ices, dust, rock and even organic compounds (think of a dirty snowball).

The exact composition of a comet or any small body gives clues about the composition of the Solar nebula where the body formed. This is another good reason to study small bodies, as they contain within them material unperturbed since the formation of the solar system[4] (Horner and Jones 2010; Altwegg et al. 2015).

In addition to their composition, the orbital properties of currently existing SSSBs give vital clues about the primordial evolution of the giant planets (Malhotra 1995; Gomes 2003; Kortenkamp et al. 2004) and can indicate the existence of unseen planets (Brown 2017; Batygin and Morbidelli 2017).

To summarize, small solar system bodies should be studied for the following reasons:

- To protect Earth
- To learn more about the formation of our solar system
- To detect unseen planets.

2.2 How Small Solar System Bodies Are Studied

Many different properties of SSSBs are of interest. These include orbital properties, composition, size, satellites, rings, surface features and even color. The main methods used to determine such properties are

- Probes
- Physical samples
- Computer simulations
- Telescopic observations.

2.2.1 Probes

Since Sputnik, the first artificial satellite, was launched in 1957, humans have been sending man made objects into space. A probe is an unmanned spacecraft designed to study its environment and then transmit that information back to scientists on Earth.

Probes have visited all eight planets in the solar system; have landed on Mars and Venus; have examined planetary rings and have plunged into the interiors of Jupiter and Saturn. In addition, probes have also been used to study asteroids, comets, moons, the dwarf planet Pluto and Ultima Thule (also known as 2014 MU69), a small body beyond the orbit of Neptune.

Probes have the advantage of being able to view the object in question close up. This allows the SSSB to be studied in more detail than it could be by using long

---

[4]https://www.mps.mpg.de/planetary-science/small-bodies-comets-research (accessed Dec. 28, 2017).

distance telescopic observations or computer simulations. Such probe missions have revealed much information about the composition, shape, formation and surface features of SSSBs.

The entire history of probe exploration of the solar system is too extensive to discuss here. Some probe missions to SSSBs and dwarf planets of note include:

- Cassini—returned a wealth of information on Saturn, its rings and moons. Its accompanying probe Huygens landed on Titan marking the first time a man made object landed on an outer solar system body.
- DAWN—visited the dwarf planet Ceres and the asteroid Vesta, revealing new information about their formation.[5]
- Deep Impact—sent an impactor into comet Tempel 1, causing an impact crater and throwing out ejecta for analysis.
- Hayabusa 2—will send an impactor into asteroid Ryugu and collect a sample of ejecta for return to Earth (ongoing).
- Lucy—will be the first space mission ever to Jupiter's Trojan asteroids. It will also visit one Main Belt Asteroid. Launch is scheduled for October 2021.[6]
- New Horizons—sent back the first high resolution photos of the dwarf planet Pluto and the small body Ultima Thule in an orbit beyond Neptune.
- OSIRIS-REx—will map the surface of the asteroid Bennu and eventually return a sample to Earth (ongoing).
- Rosetta—analyzed comet 67P/Churyumov–Gerasimenko and sent its lander, Philae, to the comet's surface.
- Stardust—collected comet debris from comet Wild 2 and transported it back to Earth.

2.2.2   Physical Samples

It is also possible to study SSSBs by examining physical samples of them here on Earth. The Stardust mission has already returned physical samples from a comet, and OSIRIS-REx and Hayabusa 2 will return samples from asteroids. Interestingly, it is not absolutely necessary to travel to a SSSB to obtain a sample.

Tons of debris in the form of dust particles from asteroids, comets and other sources enters Earth's atmosphere each day. These particles can be collected and returned to Earth by balloons or aircraft in the upper atmosphere[7] (Potashko and Viso 2014). Furthermore, since many meteorites originate from asteroids, they too are physical samples.

---

[5]https://dawn.jpl.nasa.gov/ (accessed Feb. 8, 2018).

[6]https://www.nasa.gov/content/goddard/lucy-the-first-mission-to-jupiter-s-trojans (accessed Apr. 6, 2019).

[7]https://curator.jsc.nasa.gov/dust/.

2.2.3 Computer Simulations

The use of computers has become essential in virtually all areas of astronomy, and studying SSSBs via computers is far less expensive than a probe mission.

Using computer simulations, the orbit, rotation, rings and satellites of SSSBs are just a few of the things that can be studied over time spans of hundreds, thousands, millions or even billions of years—an impossible feat for a single space probe or telescope.

Simulations of the orbital evolution of small bodies can help determine the past orbital history, likely fate and threat the body poses to Earth. Such simulations may involve two, three or even thousands of simulated bodies. A **numerical integrator** is an algorithm designed to advance a system of N interacting bodies from an initial state, Ψ_0, to a final state, Ψ_f.

The final state may be after or before the time of the initial state. The interactions may be purely gravitational or may include non-gravitational forces.

The N-body problem is the study of the dynamics of N interacting point masses. **Dynamics** is the study of the motion of objects due to forces.

Only the 2-body problem can be solved analytically for the general case (Murray and Dermott 1999; Gurfil et al. 2016). The solution to the general case of the 3-body and larger N-body problems can only be approximated numerically using a numerical integrator. The set of point masses under study is called a **system**.

Generally, the integration from the initial to the final state does not occur in one step, but instead proceeds through a series of intermediate steps. That is, the integrator advances the initial state to some intermediate state, Ψ_1, and then integrates that state to another intermediate state, Ψ_2, and so on until the integration ends, and the final state has been reached. The process of moving from one state to the next is known as **integration**. How the integration is done varies with the integrator.

The entire collection of recorded states along with the properties that govern the integration is called a **simulation**. Properties of a simulation include

- Bodies—these are the sizes and masses of the bodies that make up the system such as planets, stars, dust, small bodies and even massless bodies.
- Forces—these are the forces which drive the motion of the bodies in the simulation. These may be gravitational, collisional, drag forces and other forces.
- Initial state—includes the starting time, initial positions, and initial velocities of all bodies in the system.
- Integration Time—this is the total simulated time.
- Numerical Integrator—this is the particular integrator used for the integration, and there are many to choose from.
- Output Time—this is the time interval at which data from the simulation is recorded. Data include the state of the system and a list of bodies removed from the system along with the method of removal (collision, ejection from the solar system, etc.) and removal times. This quantity must be set carefully. Setting the output time to too small a value results in the simulation taking longer than it

should, as well as too much unneeded data. Setting the output time to too large a value may result in skipping over desired data.

- Rules—rules can encompass many things, such as under what circumstances a body is removed from the simulation, the distance between bodies at which a collision is said to have occurred, and other rules.
- Time Step—the amount of time between the consecutive states Ψ_{n-1} and Ψ_n. The time step may be constant or variable.

A group of N gravitationally interacting bodies such as SSSBs and planets, for example, can be represented as a system of point masses. The state of the n_{th} point mass, ψ_n, as shown in Eq. (2.1) can be defined as the set of components of its position vector, $\vec{r}_n$, its velocity vector, $\vec{v}_n$, and a value for time, t.

$$\psi_n = \left[\vec{r}_{nx}, \vec{r}_{ny}, \vec{r}_{nz}, \vec{v}_{nx}, \vec{v}_{ny}, \vec{v}_{nz}, t \right] \tag{2.1}$$

The combined states of all point masses that make up the system define the state of the system, Ψ, as shown in Eq. (2.2).

$$\Psi = \left[\psi_1, \psi_2 \dots \psi_N \right] \tag{2.2}$$

A variety of numerical integrators exist. One of the most basic integrators is **Euler's Method**. This method uses a constant time step, Δt, to advance from Ψ_{n-1} to Ψ_n. For one body this can be written as

$$\vec{r}_{1n} = \vec{r}_{1(n-1)} + \vec{v}_{1(n-1)} \Delta t \tag{2.3}$$

$$\vec{v}_{1n} = \vec{v}_{1(n-1)} + \vec{a}_{c1(n-1)} \Delta t \tag{2.4}$$

The time for the Ψ_n state is found from

$$t_n = t_{n-1} + \Delta t \tag{2.5}$$

Here, $\vec{a}_{c1(n-1)}$ is the acceleration vector of body 1 for the $(n-1)_{th}$ state. For the n_{th} state, $\vec{a}_{c1n}$ can be found by the instantaneous net force acting on the body and Newton's 2nd Law. Errors occur in $\vec{r}_{1n}$, $\vec{v}_{1n}$ and $\vec{a}_{c1n}$ with each integration due to the time step Δt being non-zero and the precision of the computer being used. A smaller time step reduces error but requires more time to perform the task. The experimenter must select a time step, Δt, which is small enough to achieve a desired accuracy and large enough to complete the task within an allotted time.

Other more efficient integrators reduce the error using more efficient code rather than a smaller time step. Integrators may also make use of a variable time step to improve accuracy. For example, when a small body has a close encounter with a planet, it may cause a large acceleration, which greatly increases the error when integrating. But an integrator with a variable time step shrinks the size of the time step when close encounters occur and then enlarges it again after the encounter. The result is reduced error with only a minimal increase in task time.

Other integrators include the MERCURY (Chambers 1999) collection of integrators, Bulirsch-Stöer (Hairer et al. 1993), SWIFT (Levison and Duncan 1994), and the REBOUND N-body simulation package, which is a suite of integrators including Wisdom-Holman Fast (Rein and Tamayo 2015), and IAS15 (Rein and Spiegel 2015).

The Bulirsch-Stöer Method improves on the Euler method by partitioning the time step Δt into n steps of size h.

$$h = \frac{\Delta t}{n} \tag{2.6}$$

where n can be any positive integer. The first step in advancing from Ψ_0 to Ψ_1 is the Euler method. For Body 1, the position and velocity for state Ψ_1 can be written as

$$\vec{r}_{11} = \vec{r}_{10} + \vec{v}_{10}h \tag{2.7}$$

$$\vec{v}_{11} = \vec{v}_{10} + \vec{a}_{c10}h \tag{2.8}$$

but then state Ψ_2 is found by advancing from Ψ_0 to Ψ_2 using a time step of $2h$ and vectors $\vec{a}_c$ and $\vec{v}$ from state Ψ_1. The position and velocity of Body 1 for state Ψ_2 can be written as

$$\vec{r}_{12} = \vec{r}_{10} + \vec{v}_{11}(2h) \tag{2.9}$$

$$\vec{v}_{12} = \vec{v}_{10} + \vec{a}_{c11}(2h) \tag{2.10}$$

Similarly, intermediate values $\vec{r}_{13}, \vec{r}_{14} \ldots \vec{r}_{1(n-1)}$ and $\vec{v}_{13}, \vec{v}_{14} \ldots \vec{v}_{1(n-1)}$ are found for states $\Psi_3, \Psi_4 \ldots \Psi_{(n-1)}$. The state Ψ_n is found by averaging two different estimates each for the position and velocity for the n_{th} state: one found by advancing from state $\Psi_{(n-2)}$ using a stepsize of $2h$ as before and the other found using Euler's method advancing from state $\Psi_{(n-1)}$ using a stepsize of h. The position and velocity of Body 1 at state Ψ_n are each found from the average of their two estimates

$$\vec{r}_{1n} = \frac{1}{2}\left[\vec{r}_{1(n-2)} + 2h\vec{v}_{1(n-1)} + \vec{r}_{1(n-1)} + h\vec{v}_{1(n-1)} \right] \tag{2.11}$$

$$\vec{v}_{1n} = \frac{1}{2}\left[\vec{v}_{1(n-2)} + 2h\vec{a}_{c1(n-1)} + \vec{v}_{1(n-1)} + h\vec{a}_{c1(n-1)} \right] \tag{2.12}$$

Typically, since the mass of a SSSB is much smaller than that of a planet, in simulations involving SSSBs and planets, the mass of a SSSB is ignored, and each SSSB is represented as a massless point mass called a **test particle**.

Test particles are very useful in reducing the time needed to run a simulation, because massless particles do not exert a gravitational force on other bodies in the system. They are also useful in studying the dynamical past or future of a SSSB as **clones** of the object under study.

Orbital parameters of clones of a known SSSB are normally created by evenly partitioning each desired orbital parameter across the range of possible values to create a set of possible values for each orbital parameter of the SSSB and then creating all possible combinations of all orbital parameters. The number of values for each orbital parameter is decided by the experimenter.

If n_1 is the number of values of the first orbital quantity, n_2 is the number of values of the second orbital quantity and so on, then the total number of clones created is given by

$$n_1 \times n_2 \ldots \times n_6 \tag{2.13}$$

Thus, clones of a real body are test particles that lie in orbits with orbital quantities that each lie within the uncertainty of the accepted value of that quantity.

Example Suppose a mythical small body has a semimajor axis of $a = 12.1 \pm 0.1$ au and an eccentricity of $e = 0.3 \pm 0.02$. Clones of this body could be created by partitioning each orbital parameter into three equally spaced values across the range of possible values. What would be the value of the semimajor axis and eccentricity of each clone, and how many clones would there be?

Solution This would yield three values of semimajor axis: 12, 12.1 and 12.2 au and three values of eccentricity: 0.28, 0.3 and 0.32. If other orbital parameters are held constant, then from Eq. (2.13) the total number of clones would be $3 \times 3 = 9$ clones. The semimajor axis and eccentricity of each clone are shown below. Other orbital parameters are held constant and are not shown.

a	e
12	0.28
12	0.3
12	0.32
12.1	0.28
12.1	0.3
12.1	0.32
12.2	0.28
12.2	0.3
12.2	0.32

Though error is introduced in the position and velocity vectors of each particle with each step in any integration, if a very large number of clones is used, their statistical behavior as a whole may be said to represent possible behaviors of the actual body. That is why as many clones as possible should be used.

2.2.4 Telescopic Observations

Traveling directly to an asteroid or comet can be expensive and time consuming. Much can be learned from computer simulations, but regardless of how well the algorithm is written, the object under study is still not the real thing, and properties such as color and size can't be found from test particles.

Using telescopic observations, reflected light or other types of electromagnetic radiation from a SSSB can be viewed directly with a ground-based or orbiting telescope. **Photometry** is the study of electromagnetic radiation measurement.

Photometric measurements typically consist of measurements of the intensity of electromagnetic radiation over a range of wavelengths. Photometric observations of SSSBs yield information about composition, rotation rate and other properties.

Observations from traditional optical telescopes usually yield little information about surface features on SSSBs since they are so small, however, telescopic observations can allow the motion of a SSSB relative to the stars to be tracked. This is called **astrometry**. Two organizations at the forefront of SSSB discovery via astrometry include Pan-STARRS and the Catalina Sky Survey.

Using astrometry, astronomers can determine the orbit of known SSSBs and discover previously unknown bodies. Typically, astrometry is done by taking multiple images of the same stars using a telescope that tracks the stars' motion due to Earth's rotation. Each consecutive image is taken after a given time interval. Using this method, the stars do not move relative to each other in each image. Any SSSB in the region will appear to move relative to the stars due to its orbit about the Sun and thus can be detected.

The quality of images depends on such quantities as **light pollution** and atmospheric turbulence. Light pollution degrades the signal-to-noise ratio and is caused by streetlights, neon signs and other artificial light sources.

When observing a SSSB telescopically, the altitude of the body above the horizon should be as high as possible to reduce effects of turbulence from the atmosphere. At low altitudes, atmospheric turbulence tends to degrade the quality of images due to the increased thickness of the atmosphere through which the object is viewed.

For example, a view towards an observer's horizon may be through about 100 miles of atmosphere, but a view towards an observer's zenith may be through only about 10 miles of atmosphere.

Seeing is a term used to describe the sharpness of images of celestial objects viewed through a telescope and depends on the turbulence of the atmosphere. Bad seeing results from a very turbulent atmosphere, which is notorious at low altitudes.

Over the course of 1 day, a celestial object like a star or SSSB reaches its maximum altitude when it crosses the meridian while moving west (excluding observers at the North and South Poles). The **meridian** is an imaginary curve drawn through due north, due south and the point in the sky directly above your head. For any particular celestial object, this happens only once per day. Therefore, the best time to image a celestial object during the course of 1 day is when the object crosses its meridian moving west relative to the horizon.

In addition to seeing and altitude, the orbital parameters of the SSSB in question must be taken into account for optimum viewing. SSSBs outside of Earth's orbit, in orbits of low eccentricity that are not Earth crossing, are best viewed at opposition for maximum brightness.

SSSBs inside Earth's orbit, in orbits of low eccentricity that are not Earth crossing, are best viewed at maximum elongation to remove the body from the Sun's glare as much as is possible.

Example At sunset, an observer attempts to view an asteroid as it is just above the horizon in the west. Explain why this is the best, mediocre or worst time of day to view this object. Is the asteroid in a good position in its orbit relative to Earth for prime viewing?

Solution This is worst time of day to view this object because it is so close to the horizon. The object is best viewed as it crosses the meridian while moving west relative to the horizon. The object is also best viewed when it is opposite the Sun in the sky or at opposition. Because it is so close to the Sun in the sky, it is not in a good position in its orbit relative to Earth for prime viewing.

2.2.5 The Future Study of SSSBs

In general, the future looks bright for the study of all small solar system bodies. Despite cuts at NASA, the New Horizons probe to Pluto and Ultima Thule was a huge success and allowed these bodies to be seen in high resolution for the first time.[8] Future missions OSIRIS-REx and Hayabusa 2 promise to return samples from asteroids Bennu and Ryugu.

Private citizens are now more heavily involved in solar system exploration than ever before. The launch of the Planetary Society's LightSail (and the impending LightSail 2) into Earth orbit marks the first time in history that an orbiting object of this type has been privately funded.

Citizen scientist programs continue to allow private citizens to perform real scientific research, which decades ago was only done by professionals.

Exercises

1. The comet 73P/Schwassmann-Wachmann has approached Earth within a distance of 0.0787 au. Would this object be classified as a PHO? Must the MOID of this comet's orbit be equal to 0.0787 au? Explain your answer.

---

[8]https://www.nasa.gov/mission_pages/newhorizons/main/index.html (accessed Jan. 18, 2018).

2. Explain why a large icy body entering the inner solar system could be a threat to Earth without actually colliding with it.

3. During the formation of the solar system, Earth would have been too hot for water to form. Yet today, most of Earth's surface is covered with water. Discuss a likely source of Earth's water.

4. Give one reason for studying small bodies of the solar system. Find a journal article or reputable website that backs up your reason.

5. List what advantages studying small bodies by computer simulations has over studying small bodies using probes and vice versa.

6. Explain the difference between a test particle and a small speck of dust in space.

7. Do some research or talk with a professional and make a list of N-body integration algorithms currently being used in solar system research.

8. Given an asteroid whose orbit lies outside of Earth's, has relatively low eccentricity and does not cross Earth's orbit. An observer decides to image the asteroid telescopically when it is just rising above the horizon a few minutes before sunrise. What mistakes is this observer making?

9. Explain how a sample of an asteroid could be obtained without actually going to the asteroid.

10. Given an icy body which orbits the Sun in an orbit with a semimajor axis of 15.864 ± 0.005 au and an eccentricity of 0.175 ± 0.003. Create clones of this body using three values of semimajor axis and three values of eccentricity, each evenly spaced across their uncertainty. How many total clones would this yield?

11. Suppose clones of a small body named Chariklo were created using 7 values of semimajor axis, 7 values of eccentricity and 7 values of longitude of perihelion while holding all other orbital parameters constant. How many clones would be created by forming all possible combinations of orbital parameters?

Chapter 3
A Brief Review of Cosmological Models

The ancients had no telescopes and so could only use their own eyes to observe the heavens. They saw the Sun, the Moon, the stars and the planets Mercury, Venus, Mars, Jupiter and Saturn.

They would have observed the same types of motion in the sky that we also observe in modern times. Today, we see the Moon periodically traverse through phases from New to Full. We see that the heavens seem to spin around Earth in an east-west direction relative to the horizon.

All heavenly bodies including the stars, Sun, Moon and planets move with an east-to-west motion relative to the **horizon**, an imaginary line where the sky meets the Earth. Motion relative to the horizon due to Earth's rotation is known as **diurnal motion**.

One way to understand diurnal motion is to imagine that all celestial objects are attached to a giant rotating invisible sphere called the **celestial sphere**, with Earth at its center.

The sphere pivots about two points on the sphere so that if a star is located at either point, it does not appear to move in the sky. These pivot points on the sphere are known as the **north celestial pole** (NCP) and the **south celestial pole** (SCP) and are directly above Earth's North and South Pole, respectively.

As this sphere rotates, all heavenly bodies are seen to move east-to-west relative to the horizon. Bodies can be seen to rise above the horizon and set below it. However, some stars neither rise nor set but instead are seen to spin around a celestial pole all night. These stars are known as **circumpolar stars**.

Which stars are circumpolar depends on the **latitude** of the observer. One definition of latitude is the smaller angle between the horizon and whichever celestial pole is above the horizon along the meridian. Then, all stars which are located within the latitude angle to the celestial pole are circumpolar. For northern latitudes, the latitude angle is measured to the north celestial pole, and for southern latitudes it's measured to the south celestial pole.

J. Wood, *The Dynamics of Small Solar System Bodies*, SpringerBriefs in Astronomy, https://doi.org/10.1007/978-3-030-28109-0_3

Angles on the celestial sphere can be measured in **degrees, arcminutes** or **arcseconds**. There are 60 arcseconds in an arcminute and 60 arcminutes in a degree. Angles on the sphere are very important and can be used to determine such things as the angle across a celestial object (the **angular diameter**), the angle between two celestial objects (the **angular distance**) and the day of the year.

As an example of angle measurement, the point directly above an observer on Earth on the celestial sphere makes a 90° angle to any point on the horizon. This point is called the **zenith**, and the point on the sphere directly beneath the observer is called the **nadir**.

Thus, an observer at the North or South Pole sees all visible stars as circumpolar. At these two locations, the stars move parallel to the horizon and maintain the same altitude to it. At other locations on Earth during the course of a day, a celestial body changes its altitude.

While non-circumpolar stars cross the meridian once per day and reach their maximum altitude when doing so, circumpolar stars cross the meridian twice. When crossing below a celestial pole in the sky, the circumpolar star is said to cross the lower meridian, and when crossing above a pole, the star is said to cross the upper meridian where it reaches its maximum altitude.

Other points and curves on the celestial sphere are important for determining such things as the time of day and the season of the year. The projection of Earth's equator onto the celestial sphere is known as the **celestial equator**. The Sun moves along a path known as the **ecliptic** through the 13 constellations of the **zodiac**. This curve intersects the celestial equator at an angle of 23.5° because of the tilt of Earth's rotation axis.

The celestial sphere for an observer at 40° north latitude is shown in Fig. 3.1. The observer is located at the x in the center of the gray circle, and the observer's zenith and nadir are shown. The meridian curve is shown in red. Along the curve, the north celestial pole makes a 40° angle with due north (N) on the horizon, and all stars on the celestial sphere within 40° of the NCP are circumpolar. The celestial equator makes a 40° angle south of the zenith along the meridian. Thus, another way to measure latitude is to find the angle between an observer's zenith and celestial equator along the meridian. The celestial equator is north of the zenith at southern latitudes and south of the zenith at northern latitudes.

The cardinal directions north, south, east and west are labeled as N, S, E and W along the horizon respectively. The celestial equator is shown in green and the ecliptic in yellow. The dashed lines between the two celestial poles represents Earth's axis of rotation.

Example An observer in a southern latitude 'sees' the celestial equator at an angle of 10° to the point due north on the horizon, as measured along the meridian. At what latitude is this observer?

Solution To find the answer, imagine a person standing at the equator on Earth. At this location, the celestial equator is directly overhead and makes a 90° angle to due north. Now, as the person walks south, for every degree the person walks, the celestial equator moves 1° north of the zenith along the meridian. When the celestial

Fig. 3.1 The celestial sphere for an observer at 40° north latitude. The angle between the NCP and due north (N) is the latitude. All stars within 40° of the NCP are circumpolar

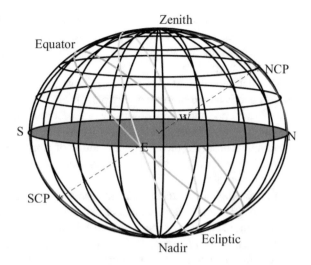

equator is 10° from the point due north on the horizon along the meridian, this means that the latitude of the observer is $90 - 10 = 80°$ south, since latitude is measured by the angle between the celestial equator and the zenith along the meridian.

Example Which stars are circumpolar for an observer on the equator?

Solution None of them. By the definition of latitude, either celestial pole would make an angle of zero to the horizon. So, there is no latitude angle to either celestial pole within which stars can be.

In addition to diurnal motion, the Sun, Moon and planets also move independently relative to the stars, while the stars themselves maintain their positions relative to each other. This means that the angle between any particular star and the Sun or Moon changes in time.

The change in angle per unit time of an object on the celestial sphere is known as **angular speed**. The Moon and Sun move west to east relative to the stars at angular speeds of about 0.5° per hour and 1° per day, respectively. Though, these speeds vary. Planets move with varying speed and direction relative to the stars. They also move relative to the Sun.

For example, the planets Mercury and Venus move relative to the Sun in the sky but never stray too far from it. Instead, these two planets bounce back and forth from some maximum angle east of the Sun to some maximum angle west of the Sun.

This maximum angle is called a **maximum elongation** and is not constant. The maximum eastern or western elongation of Venus varies between 45–47° and that of Mercury 18–28°.

But the planets Mars, Jupiter and Saturn have no maximum elongation. Instead, these planets are capable of getting a full 180° away from the Sun (or opposite the Sun) in the sky. At this point, the planet is said to be at **opposition**.

Most of the time, these planets move west to east relative to the stars at a nonuniform speed. This is called **prograde motion**. But occasionally these bodies drift westward relative to the stars in loops or zigzags before resuming their more usual eastward motion.

This occasional westward motion relative to the stars is known as **retrograde motion** and is a type of motion only planets are observed to do.

The ancients had been observing all of these motions for centuries before any attempt was made at explaining them. For example, we know that ancient Babylonians observed the retrograde motion of planets because of records found on tablets. But as far as we know, the Babylonians had no model in which the planets orbited a central body.

A **cosmological model** explains heavenly motions and accurately predicts future events. A good cosmological model explains all observed heavenly motions and predicts future celestial events to within a desired degree of accuracy. A good model is also changeable. For example, if a model fails to describe a particular motion seen in the heavens, then it is refined until it explains the motion if it's a good model or discarded if it's a bad model.

Early cosmological models explain motion without using forces. Thus, they do not correctly describe the dynamics of celestial objects the way we do today. Nevertheless, it is of interest to study such early models to gain an understanding of the scientific process.

3.1 The Geocentric Model

It was the ancient Greeks who first tried to explain the motions they were seeing in the heavens. From the Greeks point of view, everything seemed to revolve around the Earth. Having neither computers nor space probes, nor even knowledge of gravity, we can hardly blame the Greeks for inventing the erroneous geocentric (or Earth-centered) model.

In an ancient Greek **geocentric model**, Earth was the center of all revolution. To an ancient Greek, orbits did not exist as we know them today. Instead the Sun, Moon, planets and stars were each attached to an invisible crystalline sphere called a **deferent**, which constantly spun around the Earth, and the Earth itself did not rotate.

But this model proved to be too simplistic and was unable to explain why planets changed their angular speed in the sky or retrograde motion.

Over the centuries, different incarnations of the geocentric model were formulated in an attempt to align all heavenly motion with the model. In order to explain why planets seemed to speed up and slow down in their motion against the stars, Earth was moved off-center of the deferents which contained planets.

Under this model, an observer at the center of the deferent would view a planet moving at a uniform rate, however, an observer on Earth would observe the planet speeding up and slowing down as its distance to Earth changed. When the planet was closer to Earth, it appeared to move faster in the sky, and when farther away appeared to move more slowly.

Having Earth off-center was known as an **eccentric**. The stars had their own sphere centered on Earth since their motion relative to the horizon was uniform, and stars were never seen to move relative to each other. Each planet had its own distinct eccentric.

The Greek astronomer Hipparchus even devised a method to explain retrograde motion of planets while still maintaining spheres spinning in the same direction. He did this by placing the planet on a smaller spinning sphere, the center of which was fixed to the rim of a constantly spinning, much larger deferent. The smaller sphere was called an **epicycle**.

The most successful geocentric model was devised in the second century AD by the astronomer Claudius Ptolemy and today is called the **Ptolemaic model**. As explained in his book *The Almagest*, in this model the centers of the epicycles of Mercury and Venus were fixed to an Earth-Sun line. The center of the motion was a point exactly off-center opposite of Earth and was called the **equant**. This was a substantial difference compared to earlier geocentric models.

For an observer at the equant, the motion of the center of the epicycles of Mars, Jupiter and Saturn was uniform. But for an observer on Earth, the motion was not uniform. Ptolemy set the equant position; deferent size and rotation speed; epicycle size and rotation speed at the values he desired for each planet in an attempt to make his model align with observations.

The addition of the equant explained why the different retrograde motions of a planet had different durations and trajectory shapes such as zigzags and loops.

Ptolemy explained how Mars, Jupiter and Saturn were brighter and moved faster during retrograde motion than during prograde motion by forcing the radii of their epicycles to always be parallel to an Earth-Sun line.

This forced those planets to retrograde only when at opposition when they were closest to Earth and thus at their brightest. To explain why Mercury and Venus had maximum elongations and never got opposite the Sun in the sky, Ptolemy fixed the centers of their epicycles on an Earth-Sun line.

Figure 3.2 shows an example of the Ptolemaic model for one planet with its deferent, epicycle, equant and eccentric. An × marks the location of the center of the deferent.

Figure 3.3 shows the Ptolemaic model for the planets known at that time, the Sun and the Moon. Earth is at ×. The Moon and Sun orbit Earth with no epicycles. The planets Mercury, Venus, Mars, Jupiter and Saturn are shown along with their epicycles and deferents. Note how the epicycles of Mercury and Venus are fixed on an Earth-Sun line and how the radii of the epicycles of Mars, Jupiter, and Saturn are parallel to the Earth-Sun line.

Fig. 3.2 An example of the Ptolemaic model showing the deferent, eccentric, equant and epicycle for one planet. Both the deferent and epicycle spin counterclockwise

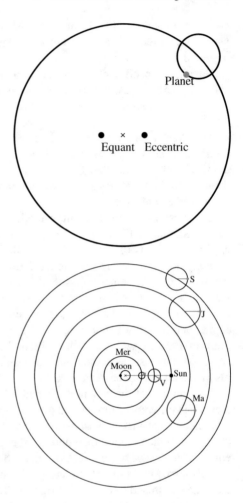

Fig. 3.3 The Ptolemaic model. Earth is at ×. Shown is the Moon, Sun and planets Mercury (Mer), Venus (V), Mars (Ma), Jupiter (J) and Saturn (S) with their epicycles and deferents

3.2   The Copernican Model

The Ptolemaic model reigned as the accepted model of the universe until the sixteenth century, when Polish astronomer Nicolaus Copernicus revived a little-known, ancient Greek model known as the **heliocentric model**. In such a model, the Moon orbits the Earth, but the planets orbit the Sun.

At that time, the heliocentric model was contrary to the teachings of the Catholic Church. Believing that Earth orbited the Sun could result in imprisonment, torture, or even death. No wonder that his book *De Revolutionibus Orbium Coelestium*, which presented the model, was only published just before his death in 1543.

In this model, Earth was replaced as the center of revolution with the Sun. Planets still resided on invisible spheres and still had epicycles though they were smaller than those in the Ptolemaic model. Heavenly motions were still explained but in a different way.

Fig. 3.4 The heliocentric
Copernican Model. The
Moon (not shown) orbits
Earth, but the planets orbit the
Sun located at the ×. The
order of the visible planets
from the Sun is Mercury,
Venus, Earth, Mars, Jupiter,
and Saturn. Only the orbits of
Mars, Jupiter and Saturn are
labeled. Epicycles are not
shown. The stars are attached
to a motionless sphere

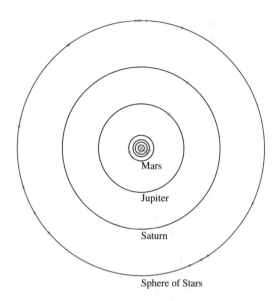

Sphere of Stars

The stars did not move at all, but resided on a motionless invisible sphere of their own at a distance from the Sun beyond any planet. No forces were involved, and orbits as we know them today still did not exist.

Copernicus was also able to determine the relative distance of each planet to the Sun compared to Earth's distance. Today we would refer to this Earth-Sun distance as an **astronomical unit** or au: 1 au $= 1.49 \times 10^8$ km. He correctly found that the order of the visible planets from the Sun was Mercury, Venus, Earth, Mars, Jupiter, and Saturn. The model is known today as the **Copernican model** and is shown in Fig. 3.4.

Unlike the Ptolemaic model, the Copernican model correctly explained diurnal motion using a rotating Earth. The Copernican model explained retrograde motion as an illusion due to the passing of planets. As a faster-moving Earth caught up and overtook a slower-moving outer planet such as Mars, it would only appear to move backwards and would not physically move backwards as it had in the Ptolemaic model.

This is shown in Fig. 3.5. As the faster moving Earth catches up to and passes by Mars, it appears to move backwards in a loop as shown. Afterwards, Mars resumes its prograde motion. The duration of the retrograde is about 50 days.

This perfectly explained why outer planets were in the middle of retrograde motion during opposition, which occurred when the Earth was between the Sun and the outer planet along a line. When the Sun was in between the outer planet and Earth, the alignment was called **conjunction**.

Inner planets could similarly align with the Earth and the Sun in two ways. If the inner planet was between the Sun and Earth along a line, then this alignment was called **inferior conjunction**, and if the Sun was in between the inner planet and the Earth, the alignment was called **superior conjunction**. Alignments between Earth, the Sun and planets are shown in Fig. 3.6.

Fig. 3.5 The retrograde
motion of Mars as viewed
from Earth. In this example,
Earth and Mars both orbit the
Sun counterclockwise, and
Mars retrogrades in a loop for
about 50 days

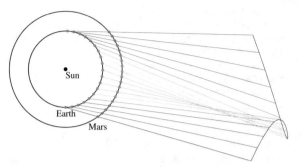

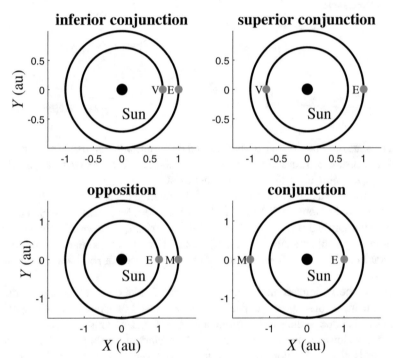

Fig. 3.6 Inferior and superior conjunction between Venus (V) and Earth (E). Opposition and conjunction between Earth and Mars (M)

The Copernican model did little to improve accuracy. Either the Ptolemaic or the Copernican model could be used to predict future celestial events with about the same accuracy, which is one of the reasons why the Copernican model was not immediately universally embraced.

However, what was significant was that the Copernican model was more physically correct than the Ptolemaic model. For example, the Ptolemaic model was only aesthetically correct in describing diurnal and retrograde motion.

This meant that it offered an explanation for these motions, but it used rotating deferents and epicycles that did not actually exist. In contrast, the Copernican model went beyond mere aesthetics and correctly described these motions using motions that actually existed, namely, a rotating Earth and revolving planets about the Sun.

Future cosmological models would improve upon the Copernican model by introducing forces as the cause of motion. Today, astronomers still seek correct physical models to explain various phenomena viewed in the heavens. As was the case with these early cosmological models, models today used to describe astronomical phenomena can also be refined or discarded in an attempt to discover the correct physical model.

Exercises

1. List the properties of a good cosmological model.
2. Use the Ptolemaic and Copernican models to explain the difference between being physically correct and being aesthetically correct.
3. List the major properties of a geocentric model according to the ancient Greeks.
4. What geometric point did Ptolemy use to explain how a planet could retrograde in different ways? What would an observer at this geometric point see when observing a planet and its epicycle?
5. In what sense is the Ptolemaic model neither geocentric nor heliocentric?
6. Explain how the Ptolemaic and Copernican models explained retrograde motion.
7. What did the planets orbit in the models of Ptolemy and Copernicus? What did the Moon orbit in each model?
8. An observer on Earth 'sees' the celestial equator always at their zenith. Where is the observer located on Earth?
9. Where on Earth could an observer stand so that there are no circumpolar stars?
10. An observer in a northern latitude 'sees' the celestial equator at an angle of 80° to the point due south on the horizon as measured along the meridian. At what latitude is this observer?

Chapter 4
Modern Orbital Mechanics

4.1 Kepler's Laws and Newton's Universal Law of Gravitation

Until the seventeenth century, it was believed that orbits could only be circular. But in 1601, German astronomer Johannes Kepler acquired the planetary positional data of Tycho Brahe, who had recently died. The data was the most accurate of its time and spanned 15 years.

Through painstaking analysis, Kepler determined that the orbits of planets were not circles but were in fact ellipses. With the introduction of ellipses, Kepler abandoned the long-held notion that planets resided on invisible spheres. Instead, he introduced for the first time in a cosmological model the idea that a force from the Sun was responsible for holding planets in their orbits. He believed erroneously that this force was magnetic in nature.

Kepler's model could predict planetary positions as much as ten times better than previous models, making it the best model up to that time. The properties of planetary orbits in Kepler's model were stated in what today are called Kepler's three laws of motion, released in 1609, 1609 and 1618.

4.1.1 Kepler's First Law: The Law of Ellipses

Kepler's first law states: *the orbit of each planet is an ellipse with the Sun at one focus. Nothing is at the other focus* (Fig. 4.1).

An ellipse is the set of points in a plane such that the sum of the distances from each point to two fixed points is a constant. Each fixed point is called a focus (and plural, foci). An example is shown in Fig. 4.2. The major axis is a line that runs through both foci and connects the sides of the ellipse. The semimajor axis, a, is half the major axis.

© The Author(s), under exclusive license to Springer Nature Switzerland AG 2019   29
J. Wood, *The Dynamics of Small Solar System Bodies*, SpringerBriefs
in Astronomy, https://doi.org/10.1007/978-3-030-28109-0_4

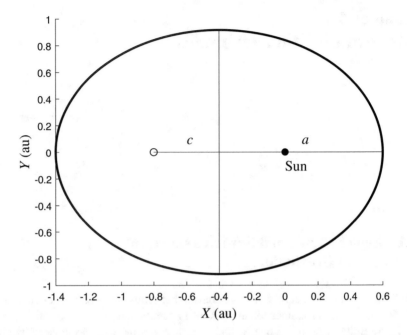

Fig. 4.1 An example of an elliptical orbit for a hypothetical planet orbiting the Sun with an eccentricity of 0.4 and semimajor axis of 1 astronomical unit—the semimajor axis of Earth's orbit around the Sun. The Sun is shown as the filled circle. The empty focus is shown as the open circle. The semimajor axis is shown as a, and the distance from a focus to the geometrical center as c

The amount by which an ellipse differs from a circle is called the eccentricity, e, of the ellipse. A circle is a special case of an ellipse for which both foci are located at the same point. A circle has an eccentricity of zero. Given c, the distance from either focus to the geometrical center of the ellipse, the eccentricity is given by:

$$e = \frac{c}{a} \tag{4.1}$$

In general, $0 \le e < 1$ for any ellipse. An orbit being an ellipse means that a planet's distance to the Sun changes as it orbits. Its distance of closest approach to the Sun is called the **perihelion distance**, q, and its farthest distance from the Sun is called the **aphelion distance**, Q. These are given by

$$q = a(1 - e) \tag{4.2}$$

$$Q = a(1 + e) \tag{4.3}$$

An example of a hypothetical planet in an elliptical orbit about the Sun is shown Fig. 4.1.

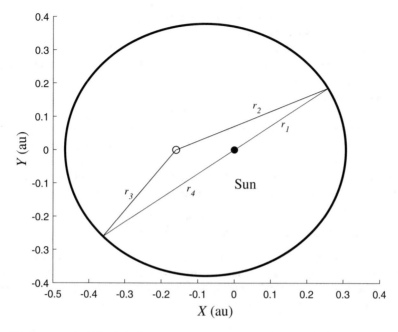

Fig. 4.2 An example of Kepler's first law for an elliptical orbit for a hypothetical planet orbiting the Sun. The Sun is shown as the filled circle. The empty focus is shown as the open circle. The distances between a certain point on the ellipse and each focus are shown as r_1 and r_2. The distances between a different point on the ellipse and each focus are shown as r_3 and r_4. For an ellipse, the following relation holds for any two points on the ellipse: $r_1 + r_2 = r_3 + r_4$

Example Derive the equation for the aphelion distance in terms of a and e using $e = \frac{c}{a}$ and the definition of the semimajor axis.

Solution The aphelion distance is along the major axis of the ellipse, and the Sun is at one focus. The aphelion distance is given by $Q =$ the distance from the point of aphelion to the center of the ellipse + the distance from the center of the ellipse to the Sun. The distance from the point of aphelion to the center of the ellipse is a definition of semimajor axis, a. The distance from the center of the ellipse to the Sun and is defined as $c = ae$. Therefore $Q = a + c = a + ae = a(1 + e)$.

4.1.2   Kepler's Second Law: The Law of Equal Areas

Kepler's second law states: *a line drawn from a planet to the Sun sweeps out equal areas in equal times.*

A consequence of Kepler's second law is that a planet varies its angular speed as it orbits the Sun. A planet moves faster nearer the Sun and slower farther away from the Sun. Thus, at its point farthest from the Sun, the planet is moving its slowest, and at its point closest to the Sun, the planet is moving its fastest. Only in a circular orbit will a planet maintain the same angular speed at all times.

A demonstration of Kepler's second law can be found in Fig. 4.3 for the planet Mercury. During an 11-day period, a line drawn from Mercury to the Sun sweeps out the shaded area as Mercury moves from point A to point B and is relatively near the Sun in its orbit.

During another 11-day period during which Mercury is relatively far from the Sun, it moves between points C and D. Again, a line drawn from Mercury to the Sun sweeps out the same area because the time to sweep out each area is the same—11 days—even though Mercury travels with different angular speeds when nearer the Sun than when farther away from the Sun. Notice that the shapes of the two shaded areas are different from each other though they are equal.

Example A line drawn from the Earth to the Sun sweeps out an area *A* in 1 week. What area does this line sweep out in a time of 8 weeks?

Solution According to Kepler's second law, a line drawn from a planet to the Sun sweeps out equal areas in equal times. Therefore, every week this line should sweep out an area *A* regardless of where the planet is in its orbit. Therefore, after 8 weeks the total area swept out would be 8*A*.

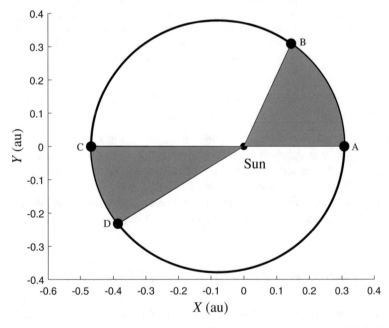

Fig. 4.3 An example of Kepler's second law for the planet Mercury. Each shaded area shown was created by a line drawn from Mercury to the Sun which swept out the area during an 11-day period. Though the shaded areas have different shapes, they are equal because the time interval during which each was created is the same

Example Given three orbits of small bodies orbiting the Sun described as follows. All three orbits have a semimajor axis of 15.0 au. The eccentricities of the orbits are: Orbit A: $e = 0$, Orbit B: $e = 0.5$ and Orbit C: $e = 0.1$. Rank these in order from left to right from smallest to largest velocity at aphelion.

Solution From the equation $Q = a(1 + e)$ we see that for a constant a, aphelion distance is linearly related to e. We know from Kepler's second law that the further the distance from the Sun the slower the body moves. So, as e increases, velocity at aphelion decreases. The velocity order from left to right from smallest to largest would be: Orbit B, Orbit C, Orbit A.

4.1.3 Kepler's Third Law: The Harmonic Law

Kepler's third law states: *the square of the orbital period of a planet, P, is directly proportional to the cube of the semimajor axis of its orbit.* The orbital period is the time it takes a planet to orbit the Sun. This can be expressed mathematically as:

$$P^2 = ka^3 \qquad (4.4)$$

Kepler's third law shows that the larger a planet's semimajor axis, the longer it takes to orbit the Sun and the slower its average orbital speed. If P is in years and a is in astronomical units, then $k = 1$ for all planets orbiting the Sun. Table 4.1 shows orbital data for the eight planets[1] and orbital period of each planet calculated using Kepler's third law. Figure 4.4 shows the linear relationship between orbital period squared and semimajor axis cubed for the eight planets.

Example Two small bodies orbit the Sun in elliptical orbits. The orbit of Body A has a semimajor axis of 2 au, and the orbit of Body B has a semimajor axis of 6 au. First, without doing any calculation, should the ratio of the orbital period of Body B to that of Body A be 3? Calculate the ratio of the orbital period of Body B to that of Body A.

Table 4.1 Orbital data for the eight planets of the solar system

Planet	a (au)	P years	e	i (°)
Mercury	0.387	0.24	0.205	7.0
Venus	0.72	0.61	0.007	3.4
Earth	1	1	0.017	0
Mars	1.52	1.87	0.094	1.9
Jupiter	5.2	11.86	0.049	1.3
Saturn	9.55	29.5	0.057	2.5
Uranus	19.2	84.1	0.046	0.8
Neptune	30.1	165	0.011	1.8

---

[1] http://ssd.jpl.nasa.gov/horizons.cgi?sbody=1#top (accessed December 31, 2015).

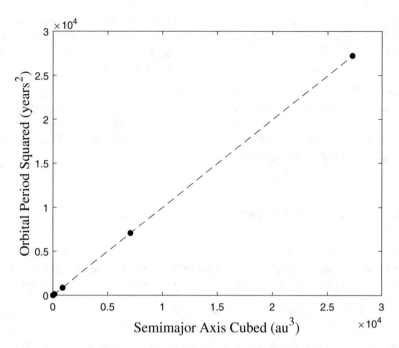

Fig. 4.4 The linear relationship between orbital period squared and semimajor axis cubed for the eight planets

Solution According to Kepler's third law, the orbital period squared is proportional to the semimajor axis cubed. Thus, orbital period is not directly proportional to the semimajor axis. Therefore, the ratio is not 3. To calculate the correct ratio, use Kepler's third law.

$$\frac{P_B^2}{P_A^2} = \frac{a_B^3}{a_A^3} = \frac{6^3}{2^3} = 27$$
$$\frac{P_B}{P_A} = \sqrt{27} = 5.20$$

4.1.4   Newton's Universal Law of Gravitation

Though Kepler originally developed his laws for planets orbiting the Sun, today we know that his laws can be applied to any elliptical orbit in the 2-body problem for which the orbiting body has negligible mass compared to the primary body.

Examples include moons orbiting planets, artificial satellites orbiting a planet, and small bodies orbiting the Sun. Though Kepler theorized that the force that held planets in orbit about the Sun was magnetic, today we know that it's actually gravity that does the job.

In the eighteenth century, Isaac Newton developed his Universal Law of Gravitation, which states that the magnitude of the gravitational force between two point particles of masses m_1 and m_2 separated by a distance d is given by:

$$F = G_c \frac{m_1 m_2}{d^2} \tag{4.5}$$

Here, $G_c = 6.673 \times 10^{-11} \frac{\text{Nm}^2}{\text{kg}^2}$ is the gravitational constant of the universe. When a small body moves only due to gravitational forces, Newton's Universal Law of Gravitation can be combined with Kepler's Laws to define orbits more quantitatively. For example, the constant k in Eq. (4.4) can be better defined if Newton's Universal Law of Gravitation is accounted for. In the case in which $m_1 \gg m_2$ Kepler's third Law becomes:

$$P^2 = \frac{4\pi^2}{G_c M_p} a^3 \tag{4.6}$$

Here, a and P are the semimajor axis and orbital period of the orbit of the small body about a central mass M_p. Thus, if a and P are known, then the mass of the more massive body can be found. For example, by finding the orbital quantities Saturn's moon Titan, the mass of Saturn can be found using Eq. (4.6).

4.1.5   Tidal Forces

Tidal forces are the result of the varying gravitational force across one body due to another body. Equation (4.5) assumes that both masses exist only at one point. However, real objects such as planets and small bodies have diameters and do not have all their mass concentrated at a single point.

As a result, when Eq. (4.5) is applied between a mass and different points on a second mass, the gravitational force is different at different points on the second mass due to the difference in position of each point.

The tidal force vector at a point on the second mass is defined as the difference between the gravitational force vector due to the first mass on a point mass at that point and the gravitational force vector due to the first mass on a point mass at the center of mass of the second mass. This can be written as: F_{tidal} = Gravitational force vector at a point − Gravitational force vector at the center of mass.

Given a small, spherical uniform body of mass m and radius r near a larger mass M_p. Using an approximation and Newton's Universal Law of Gravitation, it can be shown that the strength of the tidal force due to M_p on a point mass located at the point on the surface of the small body closest to M_p is:

$$F_{tidal} \approx \frac{2 G_c M_p m_{particle}}{d^3} r \tag{4.7}$$

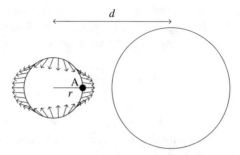

Fig. 4.5 A small body on the left is near a more massive planet on the right (not to scale). The direction of the tidal forces due to the planet at different points on the small body is shown using arrows. Note on the right side of the small body, the tidal force points toward the planet, while on the left side the tidal force points away from the planet. At point A the tidal force is given by Eq. (4.7)

where $m_{particle}$ is an infinitesimally small mass at the point in question. Figure 4.5 shows the direction of tidal forces at different points on a small body near a planet. At point A, the tidal force is given by Eq. (4.7).

Tidal forces have applications everywhere from within our own solar system to between galaxies. A spectacular demonstration of the effect of tidal forces was seen in 1994 when comet Shoemaker-Levy 9 was ripped apart into fragments by tidal forces from Jupiter and subsequently collided with the planet (Sekanina 1996).

Example Given a uniform spherical planet of radius R_p and a small moon of mass m that orbits it at a distance d from the planet's center of mass. Given point A on the planet at the point of largest distance to the moon as shown in the figure. Derive an equation for the strength of the tidal force per unit mass (the tidal acceleration) due to the moon on the planet at point A using a Maclaurin series expansion with R_p as the variable.

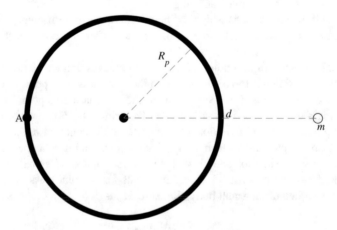

Solution F_{tidal} = the force on a particle at point A − force on a particle at the center of the planet. We need to use Newton's Universal Law of Gravitation to find these forces. But to use it, we need we need the distance from the center of the moon to Point A. Since the planet is uniform, its center of mass is also its geometrical center. Furthermore, since Point A is along a radial line from the moon, the distance from the moon to point A is $d + R_p$. Using Newton's Universal Law of Gravitation, the difference between the two forces is given by

$$F_{tidal} = G_c \frac{mm_{particle}}{(d+R_p)^2} - G_c \frac{mm_{particle}}{d^2}$$

The first term can be expanded into a Maclaurin series to yield

$$G_c \frac{mm_{particle}}{(d+R_p)^2} \approx G_c mm_{particle}\left(\frac{1}{d^2} - \frac{2}{d^3}R_p\right)$$

The tidal force is then given by

$$F_{tidal} \approx G_c mm_{particle}\left(\frac{1}{d^2} - \frac{2}{d^3}R_p - \frac{1}{d^2}\right)$$

Dividing both sides by $m_{particle}$ and canceling $\frac{1}{d^2}$ yields the tidal acceleration. The result is

$$a_{tidal} \approx -\frac{2G_c m}{d^3}R_p$$

Here, the minus sign is due to the direction of the tidal force being in the opposite direction as that of the force of gravitational attraction at the point due to the moon.

4.2   Orbits in a Three-Dimensional Coordinate System

Orbits such as the elliptical orbit described by Kepler exist in only two dimensions, however the position of a body in an orbit can be described using a traditional three-dimensional Cartesian coordinate system X, Y, and Z.

In this case, the orbit needs to be described using more than just the semimajor axis and eccentricity. Furthermore, another orbital parameter related to the position of the orbiting body as a function of time still needs to be defined.

4.2.1   The 2-Body Problem in the Solar System

In the 2-body problem in the solar system, the mass of the Sun, M_{Sun}, can be assumed to be so large compared to the masses of other solar system bodies that its motion is negligible. In this case, the force of gravity of the Sun on the orbiting body is the driving force of the motion.

For a body of mass m orbiting the Sun with position vector $\vec{r}$ from the location of the Sun, this force is given by the vector form of Newton's Universal Law of Gravitation:

$$\vec{F} = G_c \frac{mM_{Sun}}{|r|^3} \vec{r} \tag{4.8}$$

Combining this with Newton's 2nd Law $F = ma_c = m\frac{d^2\vec{r}}{dt^2}$ yields:

$$\vec{F} = m\frac{d^2\vec{r}}{dt^2} = G_c \frac{mM_{Sun}}{|r|^3} \vec{r} \tag{4.9}$$

where t is time. Canceling the mass m yields:

$$\frac{d^2\vec{r}}{dt^2} = G_c \frac{M_{Sun}}{r^3} \vec{r} \tag{4.10}$$

the solution of this equation is actually a set of curves known as conic sections. These curves are known as the ellipse (including a circle), parabola and hyperbola. The conic sections are formed by the intersection between a plane and the surface of a cone.

If the plane is parallel to the base of the cone (has a slope of zero), then the intersection forms a circle. If the slope of the plane is between the slope of the cone and zero, an ellipse is formed. If the slope of the plane matches the slope of the cone, then the intersection forms a parabola. If the slope of the plane is greater than the slope of the cone, then the intersection forms a hyperbola. The general solution for all the conic sections is

$$r = \frac{p}{1 + e\cos(\theta - \theta_o)} \tag{4.11}$$

where p is called the semi-latus rectum. θ_o is some reference angle formed by a line drawn from the Sun to some point on the curve. The angle $\theta_t = \theta - \theta_o$ is formed by the intersection of r with the major axis.

Table 4.2 shows the form of p and the restrictions on e for each conic section. For parabolic and hyperbolic orbits, the orbiting body reaches a minimum distance to the Sun, d_{min}, henceforth known as the close encounter distance.

These results can be generalized to the case in which a small body orbits a much more massive primary body. In the case of an elliptical orbit, the angle θ_t is known as the true anomaly. Figure 4.6 shows an example. A constant of any elliptical orbit is the mean motion, n_M, defined as

Table 4.2 The different conic sections with forms for p and restrictions on e for each curve

Conic section	p	e
Ellipse	$a(1 - e^2)$	$0 \le e < 1$
Parabola	$2d_{min}$	$e = 1$
Hyperbola	$a(1 - e^2)$	$e > 1$

d_{min} is the distance of closest approach between the orbiting body and the Sun. A circle is just an ellipse with $e = 0$

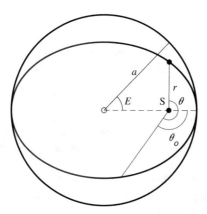

Fig. 4.6 An elliptical orbit of a body orbiting the Sun. θ_o is some reference angle formed by a reference line drawn from the Sun at point S to some point on the ellipse and the major axis. The body in orbit is shown at a distance r above point S. The angle θ is formed by the line, r, and the reference line. In the case of an ellipse, the angle $\theta_t = \theta - \theta_o$ is known as the true anomaly. The orbit is circumscribed by a circle of radius equal to the semimajor axis of the ellipse, a. The open circle is the geometrical center of the ellipse. The eccentric anomaly, E, for the orbiting body is shown. This angle will vary in time as the body orbits the Sun

$$n_M = \frac{2\pi}{P} \tag{4.12}$$

Though the true anomaly of the orbiting body does not vary linearly in time, another quantity known as the mean anomaly, M, does, and is related to the time since perihelion passage, Δt, by

$$M = n_M \Delta t \tag{4.13}$$

M has no simple geometrical definition but can be related to another angle known as the eccentric anomaly. The eccentric anomaly is defined using an elliptical orbit circumscribed in a circle of radius a.

An example is shown in Fig. 4.6. The eccentric anomaly, E, is formed by the intersection of the major axis and a radial line of the circumscribing circle that passes through a point on the circumscribing circle, which has the same horizontal (X) coordinate and same sign as the vertical (Y) coordinate as that of the position vector of the orbiting mass.

The eccentric anomaly is related to the mean anomaly via Kepler's equation

$$E - M = e\sin E \tag{4.14}$$

The orbit of a body in an elliptical orbit about the Sun is completely defined using six quantities known as the **osculating orbital parameters**, which are derived from the 3D position and velocity components of the body in orbit. These are:

- a = the semimajor axis
- e = the eccentricity
- i = the inclination, the angle between the plane of the orbit and some reference plane (often the plane of Earth's orbit about the Sun called the ecliptic plane). Values of i lie in the range $0° \leq i \leq 180°$. If $i > 90°$ the orbit is said to be retrograde.
- Ω = the longitude of ascending node, an angle in a reference plane measured between some reference line passing through the Sun and a line from the Sun to the point of ascending node, a point where the orbit intersects the reference plane and the body is moving above the reference plane.
- ω = the argument of perihelion, an angle in the plane of the ellipse between a line drawn from the point of perihelion to the Sun and a line drawn from the Sun to the point of ascending node
- M = the mean anomaly

Often, the argument of perihelion is replaced by another quantity called the longitude of perihelion, ϖ, which is the sum of the argument of perihelion and the longitude of ascending node

$$\varpi = \omega + \Omega \tag{4.15}$$

(Murray and Dermott 1999). The eccentric or true anomaly could also be used in place of M. Figure 4.7 shows the inclination, longitude of ascending node and the argument of perihelion for the case of a hypothetical planet orbiting the Sun. One other quantity of elliptical motion about the Sun is the mean longitude, λ, defined as the sum of the mean anomaly and longitude of perihelion

$$\lambda = M + \varpi \tag{4.16}$$

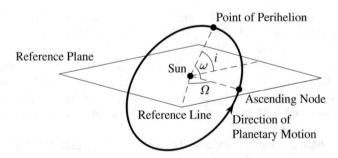

Fig. 4.7 The orbit of a planet around the Sun is shown with the planet moving counterclockwise. The square is in a reference plane. The planet intersects the plane while moving above it at the point of ascending node. The longitude of ascending node, Ω, is in the reference plane. The argument of perihelion, ω, is in the plane of the orbit. The angle between the orbital plane and reference plane is shown as i

Unlike planets, SSSBs can orbit a central body such as the Sun or a planet in parabolic or hyperbolic orbits. In this case, the central body remains at a point also called a focus. The general formula for the trajectory of an object in a hyperbolic or parabolic orbit is given by:

$$r_{radial} = \frac{h_h^2}{\mu} \frac{1}{1 + e\cos(\theta - \theta_o)} \tag{4.17}$$

where r_{radial} is the radial position of the small body from the central body located at a focus. θ_o is chosen in such a manner that the minimum approach distance occurs when $\theta - \theta_o = 0$.

Here, h_h is the constant angular momentum per unit mass of the small body (angular momentum/small body mass) and $\mu = G_c(M_p + m_s)$, where m_s is the mass of the small body.

For parabolic and hyperbolic orbits, a convenient constant of the orbit is the velocity at infinity of the SSSB, v_∞. v_∞ is undefined for an elliptical orbit. For a parabolic orbit, $v_\infty = 0$. For a given central body and small body mass, μ is constant, and a parabolic orbit is completely defined if d_{min} is known. Then h_h is given by:

$$h_h = \sqrt{2\mu d_{min}} \tag{4.18}$$

The velocity of an object moving in a parabolic orbit as a function of radial position is given by:

$$v = \sqrt{\frac{2\mu}{r_{radial}}} \tag{4.19}$$

For a hyperbolic orbit $v_\infty > 0$. For any hyperbolic orbit, the following relation applies:

$$d_{min} v_\infty^2 = \mu(e - 1) \tag{4.20}$$

Thus, if d_{min}, μ and v_∞ are known, e can be found. Then, substituting d_{min} in for r_{radial} in Eq. (4.17) yields:

$$d_{min} = \frac{h_h^2}{\mu} \frac{1}{1 + e} \tag{4.21}$$

from which h_h can be found. For any given r_{radial}, the velocity can therefore be found from:

$$v = \sqrt{\mu\left(\frac{2}{r_{radial}} + \frac{v_\infty^2}{\mu}\right)} \tag{4.22}$$

Example Given three comets that have a closer encounter with the planet Jupiter. Shown is the velocity at infinity for the orbit of each comet relative to Jupiter. State the type of orbit for each velocity at infinity.

v_∞ (km/s)	Orbit type
Undefined	
0	
3	

Solution By definition, an orbit is parabolic if $v_\infty = 0$, hyperbolic if $v_\infty > 0$ and elliptical if v_∞ is undefined. The answers are then

v_∞ (km/s)	Orbit type
Undefined	Elliptical
0	Parabolic
3	Hyperbolic

4.2.2   The Circular Restricted 3-Body Problem

In the circular restricted 3-body problem, the 3-body problem is simplified by forcing two of the bodies m_1 and m_2 to orbit their common center of mass in circular orbits. The third body is considered negligible and so does not affect the orbits of the other two masses.

Consider the case where m_1 and m_2 are a planet and the Sun. In this case, the Sun's motion can be ignored. In the rotating frame of the planet, there exists five points in the plane of the planet's orbit at which the 3rd body would feel no forces if placed there (in other words, the 3rd body would seem to be motionless to an observer on the planet). These points are known as Lagrange points and are shown in Fig. 4.8. Three of the points are located on a line that passes through the Sun and planet. These are the collinear Lagrange points and are called L_1, L_2 and L_3. L_1 lies in between the planet and the Sun. L_2 lies outside the orbit of the planet and L_3 lies on the opposite side of the Sun in the orbit of the planet.

The other two are called triangular Lagrange points and are located at points on the planet's orbit at 60° ahead and 60° behind the orbiting planet. The leading point is called L_4, and the trailing point L_5 (Murray and Dermott 1999). A body of negligible mass at any Lagrange point with no velocity in the rotating frame of the planet, $v_{rot} = 0$, will have the same mean motion, n_M, as the planet.

The distance between the planet and L_2 approximately defines the radius of a sphere centered on the planet within which the planet's gravity dominates over the Sun's. This sphere is known as the Hill Sphere (Murray and Dermott 1999), and its radius the **Hill Radius**, R_H. All known satellites of the planets orbit within their

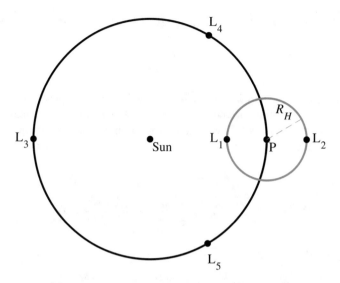

Fig. 4.8 The five Lagrange points for a planet moving counterclockwise in a circular orbit about the Sun. The planet is at point P. The Hill sphere around the planet is shown and has a radius R_H

planet's respective Hill Sphere. An equation for the radius of the Hill Sphere can be derived using the first condition of equilibrium on an object of mass m at L_2 motionless in the rotating frame of the planet.

Let the radius of the planet's orbit be r_p, and the mass of the planet be M_p. Since the planet moves in a circular orbit in the lab frame, the centripetal force is supplied by the force of gravity from the Sun. This can be expressed as

$$\frac{M_p v_p^2}{r_p} = G_c \frac{M_p M_{Sun}}{r_p^2} \tag{4.23}$$

where v_p is the magnitude of the velocity of the planet. Canceling M_p yields

$$\frac{v_p^2}{r_p} = G_c \frac{M_{Sun}}{r_p^2} \tag{4.24}$$

For a circular orbit, v_p is constant and is given by $v_p = n_M r_p$. Substituting in for v_p yields

$$\frac{n_M^2 r_p^2}{r_p} = G_c \frac{M_{Sun}}{r_p^2} \tag{4.25}$$

which can be solved for the squared mean motion. The result is

$$n_M^2 = G_c \frac{M_{Sun}}{r_p^3} \tag{4.26}$$

Let $\vec{F}_{Sun}$ be the force of gravity of the Sun on the mass m, $\vec{F}_p$ the force of gravity of the planet on the mass m and $\vec{F}_{cent}$ the centrifugal force on the mass m in the rotating frame of the planet. The equation for the first condition of equilibrium for the mass m is then written as

$$\vec{F}_{cent} + \vec{F}_{Sun} + \vec{F}_p = 0 \tag{4.27}$$

$$mn_M^2(r_p + R_H) - G_c \frac{mM_{Sun}}{(r_p + R_H)^2} - G_c \frac{mM_p}{R_H^2} = 0 \tag{4.28}$$

Canceling m yields

$$n_M^2(r_p + R_H) - G_c \frac{M_{Sun}}{(r_p + R_H)^2} - G_c \frac{M_p}{R_H^2} = 0 \tag{4.29}$$

The expression for n_M^2 found in Eq. (4.26) can be substituted into the equilibrium equation. The result is

$$G_c \frac{M_{Sun}}{r_p^3}(r_p + R_H) - G_c \frac{M_{Sun}}{(r_p + R_H)^2} - G_c \frac{M_p}{R_H^2} = 0 \tag{4.30}$$

Canceling G_c and multiplying every term by the common denominator yields

$$M_{Sun}(r_p + R_H)^3 R_H^2 - M_{Sun} r_p^3 R_H^2 - M_p r_p^3(r_p + R_H)^2 = 0 \tag{4.31}$$

This can be rearranged and the cubic expanded to yield

$$M_p r_p^3(r_p + R_H)^2 = M_{Sun} R_H^2 (r_p^3 + 3r_p^2 R_H + 3r_p R_H^2 + R_H^3) - M_{Sun} r_p^3 R_H^2 \tag{4.32}$$

and this can be simplified to

$$M_p r_p^3(r_p + R_H)^2 = M_{Sun} R_H^3 (3r_p^2 + 3r_p R_H + R_H^2) \tag{4.33}$$

in the realm where $r_p \gg R_H$ this simplifies to

$$M_p r_p^5 = 3 M_{Sun} R_H^3 r_p^2 \tag{4.34}$$

Solving for R_H yields

$$R_H = r_p \left(\frac{M_p}{3M_{Sun}}\right)^{\frac{1}{3}} \tag{4.35}$$

(Hill 1878). For elliptical orbits with low eccentricity, r_p is approximately the semimajor axis of the planet's orbit, a_p. Substituting in for r_p yields

$$R_H = a_p (\frac{M_p}{3M_{Sun}})^{\frac{1}{3}} \qquad (4.36)$$

Example Given that the Sun has a mass of 1047 Jupiter masses and Jupiter has a semimajor axis of 5.2 au, find the Hill radius of the planet Jupiter with respect to the Sun.

Solution Using Eq. (4.36) the result is

$R_H = a_p (\frac{M_p}{3M_{Sun}})^{\frac{1}{3}}$
$R_H = 5.2 (\frac{1}{3 \times 1047})^{\frac{1}{3}}$
$R_H = 0.355$ au

Though the motion of the 3rd body cannot be solved for analytically in the circular restricted 3-body problem, there are constants of the motion. In the rotating frame of the planet, the planet is taken to be at rest while the 3rd body moves due to the gravitational forces of the planet and the Sun, Coriolis force and centrifugal force. Given a 2D rectangular coordinate system, the following quantity is constant

$$C_J = n_M^2(r^2) + G_c(\frac{m_1}{r_1} + \frac{m_2}{r_2}) - v_{rot}^2 \qquad (4.37)$$

and is known as the Jacobi constant. The units are chosen so that the distance between the Sun and planet is a constant 1 and $G_c(m_1 + m_2) = 1$. n_M is the mean motion of the planet, r_1 is the separation between the Sun and the 3rd body, r_2 is the separation between the planet and the 3rd body, and r is the magnitude of the position vector of the 3rd body.

A useful application of Eq. (4.37) is to solve it for the case when $v_{rot} = 0$ for a constant value of C_J. The resulting solutions are known as zero velocity curves. These three-dimensional curves constrain the motion of a small body by forbidding it to cross them. Figure 4.9 shows examples of two-dimensional slices of zero velocity curves for six values of C_J for a hypothetical system for which the planet-Sun mass ratio is 0.2 and the planet is in a circular orbit about the Sun with a semimajor axis of 5.2 au. The X and Y axes are shown, and the Z axis points out of the page.

In the figure, a small body within the shaded region is bound to it and can never leave the region. A small body bound to a region that contains either the L_4 or L_5 Lagrange point and no other Lagrange points is said to be in a **tadpole orbit** because of the shape of its trajectory relative to the planet. This would occur for example for values of C_J of 3.5, 3.6 or 3.7. Note the tadpole-like shape of the shaded regions for the cases of these values of C_J.

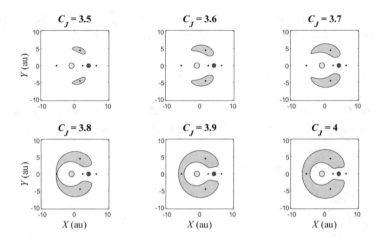

Fig. 4.9 Examples of zero velocity curves for six values of C_J for a hypothetical system for which the planet-Sun mass ratio is 0.2 in the circular restricted 3-body problem. In each diagram, the Lagrange points are shown as black dots, the Sun is the circle at the origin, and the planet is the circle on the horizontal axis on the right. A small body in the gray region is bound to that region, and a small body outside the gray region may not enter it

However, if C_J is 3.9 or 4, then the small body may librate about the L$_3$ Lagrange point. If it does, it is said to be in a **horseshoe orbit**. As can be seen in the figure, the shaded area for these C_J values resembles a horseshoe. In any diagram, a small body outside a gray region may not enter it, and a body in a gray region may not leave it.

The Jacobi constant can be rewritten in terms of osculating orbital quantities. In this form, it is called the Tisserand parameter, T_p, and is given by the Tisserand relation:

$$T_p = \frac{a_p}{a} + 2\cos(i - i_p)\sqrt{\frac{a}{a_p}(1 - e^2)} \tag{4.38}$$

Here, i is the inclination of the small body's elliptical orbit, a_p is the semimajor axis of the planet's elliptical orbit, and i_p the inclination of the planet's elliptical orbit (e.g. Levison 1996; Murray and Dermott 1999; Bailey and Malhotra 2009).

Example The orbit of comet Halley has an eccentricity of $e = 0.967$, a semimajor axis of $a = 17.83$ au and an inclination $i = 162.26°$. Given the inclination and semimajor axis of Neptune's orbit are $1.8°$ and 30.1 au respectively, find the Tisserand parameter of comet Halley with respect to Neptune.

Solution Using the equation for the Tisserand parameter yields

$$T_p = \frac{30.1}{17.83} + 2\cos(162.26 - 1.8)\sqrt{\frac{17.83}{30.1}(1 - 0.967^2)} = 1.32$$

4.3 Resonances

In the general N-body unrestricted problem in the solar system, the Jacobi constant is no longer constant, and small bodies are no longer confined by zero velocity curves. The osculating orbital parameters of small bodies can be perturbed due to collisions, gravitational forces and other forces. Gravitational forces from planets in particular can play an integral role in the dynamical behavior of small bodies over time and cause them to collide with the Sun or a planet or be ejected from the solar system.

Sometimes a special condition occurs in which cyclical gravitational perturbations from a planet on a small body combine without canceling over a sequence of conjunctions. This condition is known as **orbital resonance**. Over time scales of thousands or even millions of years, these gravitational perturbations have a "noticeable" effect on the osculating orbital parameters of the small body. Here, "noticeable" can be defined by the researcher.

The situation is similar to a swinging pendulum that is pushed every time it reaches an amplitude. Even if a single push is so small that it has no noticeable effect on the pendulum's swing, over time, the pushes add up to a noticeable increase in the amplitude of the pendulum.

Resonances exist throughout the solar system and are responsible for a variety of phenomena such as the transport of meteoroids to Earth (Wisdom 1982); the capture of small bodies into resonances during planetary migration while the solar system was still forming (Gomes et al. 2005); and the creation of gaps in small body populations or rings (e.g. Murray and Dermott 1999; French et al. 2016).

In a general sense, two bodies are in resonance with each other when the rates of change of any two orbital parameters of the two bodies exist in a ratio of two small integers. This condition is known as **orbit-orbit coupling**.

The two orbital parameters may be any combination among mean anomaly, longitude of perihelion, longitude of ascending node, etc. An object may also be in a resonance if its own spin rate and orbital period exist in a ratio of two small integers. For example, the ratio of the orbital period of the planet Mercury to its rotation period is three to two. This condition is known as **spin-orbit coupling**.

Regardless of the resonance type, whenever any two quantities exist in a ratio of two small integers, they are said to be **commensurate** with each other. Different types of resonances exist based on which orbital parameters are commensurate.

Thus, it could be said that two bodies are in resonance with each other whenever the rates of change of any two orbital parameters between the two orbits are commensurate. However, in practice this definition is too simplistic, as it does not account for the situation in which two bodies are "nearly" in resonance but the same effects caused by being in resonance are present.

In this section, the different types of resonances will be defined, and a more exact definition of resonance will be found.

4.3.1   Mean Motion Resonances

If the rates of change of the mean anomalies of two orbiting bodies are commensurate, then this is a special condition known as a **mean motion resonance** (Murray and Dermott 1999) or MMR.

From Eqs. (4.12) and (4.13) it can be seen that the rate of change of a mean anomaly is related to the inverse orbital period. Thus, it could be said that a mean motion resonance between two bodies exists whenever their orbital periods are commensurate.

Mean motion resonances may occur between a planet and a small body, a small body and ring particles or even between two planets. For example, the planet Neptune has an orbital period of 165 years, and the planet Uranus has an orbital period of 84.1 years. Taking the ratio of these two periods yields

$$\frac{165}{84.1} \approx \frac{2}{1} \tag{4.39}$$

In reality, Neptune and Uranus are nearly but not exactly in a two-to-one mean motion resonance. For this work, mean motion resonances located outside a planet's semimajor axis are written with the smaller integer first and are called exterior mean motion resonances. Mean motion resonances located inside a planet's semimajor axis are written with the larger integer first and are called interior mean motion resonances.

Given two positive integers j_1 and j_2, the notation $j_2 : j_1$ can be used to describe interior MMR when $j_2 > j_1$ and exterior MMR when $j_2 < j_1$. The special case of $j_2 = j_1$ will be discussed later.

So, Neptune is said to be nearly in a 2:1 mean motion resonance with Uranus. As another example, Fig. 4.10 shows the planet Jupiter in a 2:1 mean motion resonance with a small body. Note how Jupiter orbits the Sun once in the same time the small body orbits the Sun twice.

For Jupiter then, the 1:2 mean motion resonance is located outside Jupiter's orbit, but the 2:1 mean motion resonance is located within Jupiter's orbit. For a given MMR of a planet, the strength of the resonance is related to the **order of the resonance**, q_q, given by:

$$q_q = |j_2 - j_1| \tag{4.40}$$

(e.g. Gallardo 2006). Resonances are generally referred to as first order for $q_q = 1$, second order for $q_q = 2$ and so on.

For example, a 5 to 2 resonance would be a third order resonance since $5 - 2 = 3$. Generally, for a given MMR of a planet, a first order resonance is usually but not always stronger than a second order resonance for small integers. The location of mean motion resonances can be found using the commensurability of the orbital periods and Kepler's third law.

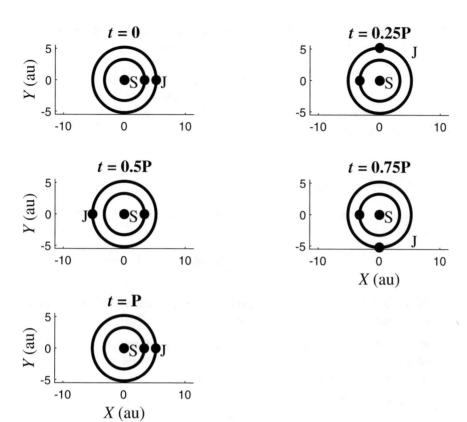

Fig. 4.10 The 2:1 mean motion resonance between Jupiter and a small body with the Sun at point S. Jupiter's orbit is labeled J. Starting with the small body in inferior conjunction with Jupiter, the two bodies are shown for different times, t, stated as fractions of P—the orbital period of Jupiter. For simplicity, the orbits are taken to be circular

Given a planet of mass M_p and a small body of negligible mass m orbiting the Sun with orbital periods P_1 and P_2 and semimajor axes a_p and a_{MMR} respectively. If the orbital periods of the two bodies are commensurate, then the integer ratio of the orbital periods can be expressed as

$$\frac{P_1}{P_2} = \frac{j_1}{j_2} \tag{4.41}$$

Equation (4.6) yields

$$P^2 = \frac{4\pi^2}{G_c M_{Sun}} a^3 \tag{4.42}$$

Squaring the integer ratio in Eq. (4.41) and setting that equal to the ratio of the square orbital periods expressed using Eq. (4.42) yields

$$(\frac{j_1}{j_2})^2 = \frac{4\pi^2}{G_cM_{Sun}} \frac{G_cM_{Sun}}{4\pi^2} \frac{a_p^3}{a_{MMR}^3} \tag{4.43}$$

and simplifying yields

$$(\frac{j_1}{j_2})^2 = \frac{a_p^3}{a_{MMR}^3} \tag{4.44}$$

which can be solved for a_{MMR}—the location of the mean motion resonance. The result is

$$a_{MMR} = a_p(\frac{j_2}{j_1})^{(2/3)} \tag{4.45}$$

Example Find the order and the location of the Neptune 9:5, Saturn 3:8 and Jupiter 1:7 mean motion resonances using the planetary data from Table 4.1.

Solution The order of each resonance can be found by subtraction: $9 - 5 = 4$, $8 - 3 = 5$, and $7 - 1 = 6$ for each resonance, respectively. The locations can be found using the equation $a_{MMR} = a_p(\frac{j_2}{j_1})^{(2/3)}$. The results are

$a_{Nep9:5} = 30.1(\frac{5}{9})^{(2/3)} = 20.3$ au
$a_{Sat3:8} = 9.55(\frac{8}{3})^{(2/3)} = 18.4$ au
$a_{Jup1:7} = 5.2(\frac{7}{1})^{(2/3)} = 19.0$ au

Figure 4.11 shows a few of the mean motion resonances of the planet Jupiter found using Eq. (4.45) and Jupiter's semimajor axis taken from Table 4.1.

When a small body is locked in a mean motion resonance with a planet, the osculating orbital elements of the small body's orbit may oscillate quasi-periodically in time with periods between 0 and 10^3 years. As an example, Moons and Morbidelli (1995) found that the semimajor axis, eccentricity and inclination of the orbits of small bodies in the 4:1, 3:1, 5:2, and 7:3 interior MMRs of Jupiter often oscillated quasi-periodically with periods on time scales of 10^3 years.

Typically, the variation of the semimajor axis in time can be described as a mixture of different harmonics due to periodic conjunctions with the planets and other harmonics.

In the simplistic, circular-restricted 3-body problem, the variation of the semimajor axis can be sinusoidal in time barring close encounters. In this case, the variation of the semimajor axis of the small body's orbit is given by

$$a = A_{proper} + A_{max}\cos(\frac{2\pi}{P}t) \tag{4.46}$$

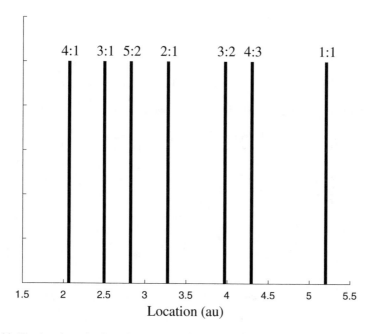

Fig. 4.11 The location of a few of the mean motion resonances of the planet Jupiter. Jupiter's semimajor axis is located at the 1:1 resonance

This equation shows that the semimajor axis oscillates in time, t, with a period P about a constant value, A_{proper}, with an amplitude A_{max}. A_{proper} is known as the proper semimajor axis. Analogously there is also a proper eccentricity and inclination.

Proper values are used to identify groups of asteroids that are members of the same family. **Asteroid families** are groups of asteroids in orbits with the same or nearly the same proper values. Families typically consist of a parent asteroid and fragments broken off in a collision.

For example, the Hungaria family, named after its parent body 434 Hungaria, consists of asteroids with osculating semimajor axes in the range 1.78–2.00 au. An asteroid family has been found among the bodies locked in the 1:1 MMR with Jupiter (De Luise et al. 2010) and among the bodies that orbit the Sun beyond the orbit of Neptune (Rabinowitz et al. 2006; Thirouin et al. 2016).

The bottom diagram of Fig. 4.12 shows an example of an oscillating semimajor axis for a test particle in the 2:1 mean motion resonance of Neptune in the plane of Neptune's orbit in the circular-restricted 3-body problem.

In the more realistic N-body problem, the oscillation of the semimajor axis is not sinusoidal and is only quasi-periodic. The top diagram of Fig. 4.12 shows how the semimajor axis of the same test particle changes in time in the 6-body problem (Sun, four giant planets, test particle).

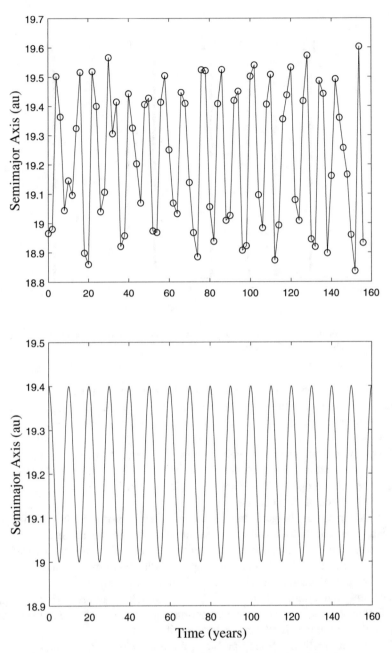

Fig. 4.12 The semimajor axis of a single test particle in the 2:1 MMR of Neptune varies in time due to gravitational perturbations. Top—in the 6-body problem. Bottom—in the 3-body problem

The perturbing force of a planet on the orbit of a test particle is related to the radial distance between the test particle and the planet via Newton's Universal Law of Gravitation, shown in Eq. (4.8).

In a system of N bodies, let $\vec{r}_{ij}$ be the radial vector drawn from the i_{th} test particle to the j_{th} planet. This displacement vector varies in time as both the planet and test particle orbit the Sun. Thus, the force perturbing the orbit of the test particle and the acceleration caused by the perturbation, a_{cij}, also vary in time.

The same argument can be made for any other planet in resonance with the test particle. a_{cij} can be expressed as an infinite sum of cosines of special angles called **resonance angles**, ϕ. This is shown in Eq. (4.47) for the j_{th} planet and the i_{th} test particle.

$$a_{cij} = \sum_z A_{cmaxz} \cos(\phi_z) \tag{4.47}$$

A resonance angle is a linear sum of the mean longitudes, longitudes of ascending node and longitudes of perihelion of the planet and the test particle.

This infinite sum is known as the disturbing function and has the general form:

$$a_{cip} \sim e^{|p_5|} e_p^{|p_6|} \cos(p_1\lambda + p_2\lambda_p + p_3\Omega + p_4\Omega_p + p_5\varpi + p_6\varpi_p) + ... \tag{4.48}$$

where p_1, p_2, p_3, p_4, p_5 and p_6 are integers subject to the D'Alembert rule:

$$p_1 + p_2 + p_3 + p_4 + p_5 + p_6 = 0 \tag{4.49}$$

(Morbidelli 2002). λ is the mean longitude of the i_{th} test particle, λ_p is the mean longitude of the j_{th} planet, Ω is the longitude of ascending node of the i_{th} test particle, Ω_p is the longitude of ascending node of the j_{th} planet, ϖ is the longitude of perihelion of the i_{th} test particle, and ϖ_p is the longitude of perihelion of the j_{th} planet.

Each term in the infinite sum is associated with a particular **subresonance** of the MMR. The MMR is defined by the values of $j_1 = |p_1|$ and $j_2 = |p_2|$ (Murray and Dermott 1999; Ellis and Murray 2000; Laskar, and Boué 2010). Each different subresonance of an MMR is defined by the unique combination of p_3, p_4, p_5 and p_6 associated with the resonance.

The term associated with the subresonance that the test particle is in dominates over other terms that tend to time average to zero. This makes a_{cij} dependent mostly on the dominant term.

When the inclination and eccentricity of the test particle are much larger than that of the planet's orbit, terms involving Ω_p and ϖ_p can be ignored. In such cases the coefficient $e^{|p_5|} e_p^{|p_6|}$ in Eq. (4.48) reduces to $e^{|p_5|} = e^{qq}$ because $e_p^{|p_6|}$ can be ignored. The resonance width is proportional to the square root of this coefficient.

Resonance angles for such cases have the format:

$$j_2\lambda_p - j_1\lambda - q_q\Omega\upsilon \tag{4.50}$$

or

$$j_2\lambda_p - j_1\lambda - q_q\varpi \tag{4.51}$$

For example, here are two possible conditions for a test particle to be in a 2:1 mean motion resonance with a planet:

$$a_{cij} \sim \cos(2\lambda_p - \lambda - \varpi) \tag{4.52}$$

and

$$a_{cij} \sim \cos(2\lambda_p - \lambda - \Omega) \tag{4.53}$$

(e.g. Roig et al. 2002; Masaki et al. 2003; Bailey and Malhotra 2009; Tiscareno and Malhotra 2009). The subresonance of the 2:1 MMR associated with the angle in Eq. (4.52) is defined by $p_1 = -1$, $p_2 = 2$, $p_3 = 0$, $p_4 = 0$, $p_5 = -1$, and $p_6 = 0$. Similarly for Eq. (4.53), $p_1 = -1$, $p_2 = 2$, $p_3 = -1$, $p_4 = 0$, $p_5 = 0$, and $p_6 = 0$.

Thus, the resonance angle for each subresonance would be $\phi = 2\lambda_p - \lambda - \varpi$ and $\phi = 2\lambda_p - \lambda - \Omega$. When ϕ oscillates quasi periodically in time, or **librates**, it means that the longitude (or angle) of the q_{qth} conjunction between the test particle and the planet changes very slowly or librates about a constant value. The resonance angle librating in time then is the definitive sign that the test particle is in a mean motion resonance with the planet (Malhotra 1994).

If ϕ does not librate but instead **circulates**—repeatedly flowing from 0° to 180° to 360° in the same direction—then $cos(\phi)$ as shown in Eq. (4.47) time averages to zero, and the associated term is not the dominant term in the disturbing function. This means that the small body is not in the $j_2 : j_1$ resonance with the planet when all associated resonance angles circulate (e.g. Murray and Dermott 1999; Roig et al. 2002; Smirnov and Shevchenko 2013).

An example of a circulating and a librating resonance angle is shown in Fig. 4.13. In the top diagram, the resonance angle $\phi = 2\lambda_N - \lambda_U - \varpi_U$ of the 2:1 MMR between Neptune and Uranus circulates. This shows that Uranus and Neptune are not locked in a 2:1 mean motion resonance.

In the bottom diagram, the resonance angle $\phi = \lambda_U - \lambda$ for a test particle locked the 1:1 MMR of Uranus librates. Here, λ_N and λ_U represent the mean longitude of the orbits of Neptune and Uranus respectively and ϖ_U is the longitude of perihelion of the orbit of Uranus.

A test particle in or out of a resonance can represent the possible behavior of a SSSB. While a small body is in a mean motion resonance with a planet, the relative severity of the perturbation on the small body's orbit at each q_{qth} conjunction is related to the relative separation of the two bodies at the time of the conjunction.

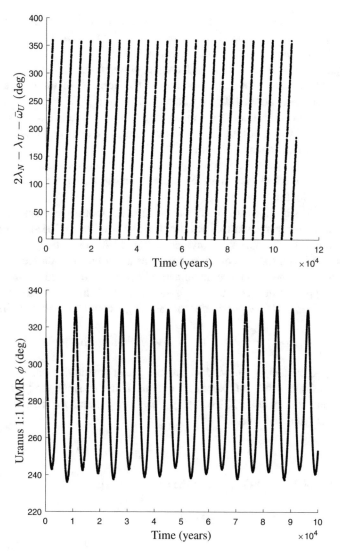

Fig. 4.13 Top—the resonance angle for the 2:1 mean motion resonance between Neptune and Uranus circulates. Bottom—the resonance angle for a test particle locked in the 1:1 mean motion resonance of Uranus librates

If the orbit of the small body crosses the orbit of the planet, then close encounters between the planet and small body are possible, and these tend to destabilize the orbit of the small body (Holman and Wisdom 1993; Duncan et al. 1995).

If a small body in a mean motion resonance is not in a planet-crossing orbit, then another effect may be that the eccentricity of the small body's orbit can be pumped up until it becomes planet crossing (Wisdom 1982).

It is also possible for no close encounters to occur even if the resonant small body crosses the orbit of the planet. Such is the case for Pluto, which is in a 2:3 exterior mean motion resonance of Neptune (Jewitt and Luu 1996).

Though Pluto crosses the orbit of Neptune, close encounters are forbidden from occurring when Pluto is at perihelion due to the libration of the resonance angle $\phi = 3\lambda - 2\lambda_N - \overline{\omega}$ about 180° (Williams and Benson 1971). Here, λ is the mean longitude of Pluto's orbit, λ_N is the mean longitude of Neptune's orbit, and $\overline{\omega}$ is the longitude of perihelion of Pluto's orbit. This results in a relatively more stable orbit for Pluto in the resonance (Malhotra 1994).

Small bodies may also become temporarily stuck in a resonance, leave it and then return. This is an effect known as **resonance sticking** (Lykawka and Mukai 2007; Bailey and Malhotra 2009).

It is also possible for three bodies to be in mean motion resonances with each other in **three-body resonances** (Smirnov and Shevchenko 2013). For example, Jupiter's moons Io, Europa and Ganymede are in a 1:2:4 three-body resonance with each other. This is a special condition known as a **Laplace resonance**.

In such cases, the resonance angle can be extended to include a 3rd body. For example, suppose that Neptune and Uranus were both in resonance with a small solar system body. Then a possible resonance angle could be written in the form

$$\phi = p_1\lambda_N + p_2\lambda_U + p_3\lambda + p_4\overline{\omega}_N + p_5\overline{\omega}_U + p_6\overline{\omega} \tag{4.54}$$

where λ_N, λ_U, λ, $\overline{\omega}_N$, $\overline{\omega}_U$, and $\overline{\omega}$ are the mean longitudes and longitudes of perihelion of the orbits of Neptune, Uranus and the small body respectively (Smirnov and Shevchenko 2013). p_1, p_2, p_3, p_4, p_5, and p_6 are integers which satisfy the D'Alembert rule:

$$p_1 + p_2 + p_3 + p_4 + p_5 + p_6 = 0 \tag{4.55}$$

(Morbidelli 2002). Then the order of the resonance is given by:

$$q_q = |p_1 + p_2 + p_3| \tag{4.56}$$

4.3.2   Secular Resonances

Other types of resonances besides mean motion resonances also exist. When the precession rates of two orbits are commensurate, the two orbits are in a state of secular resonance with each other. These commensurate precession rates may be between the precession rates of two longitudes of perihelion, two longitudes of ascending node and other pairs of orbital precession rates.

Secular resonances also perturb the orbits of small bodies but generally on longer time scales than those of MMR (Froeschle and Morbidelli 1994; Moons and

Morbidelli 1995; Murray and Dermott 1999). Some of the notable frequencies of orbital precession associated with the Jovian planets are:

- g_5 = frequency of precession of the longitude of perihelion of the orbit of Jupiter
- g_6 = frequency of precession of the longitude of perihelion of the orbit of Saturn
- g_7 = frequency of precession of the longitude of perihelion of the orbit of Uranus
- g_8 = frequency of precession of the longitude of perihelion of the orbit of Neptune
- s_5 = frequency of precession of the longitude of ascending node of the orbit of Jupiter
- s_6 = frequency of precession of the longitude of ascending node of the orbit of Saturn
- s_7 = frequency of precession of the longitude of ascending node of the orbit of Uranus
- s_8 = frequency of precession of the longitude of ascending node of the orbit of Neptune

Values of these are shown in Table 4.3. These are known as the eigenfrequencies of the solar system for the Jovian planets. The terrestrial planets have similarly defined eigenfrequencies (g_1 to g_4 and s_1 to s_4). If the precession rate of the longitude of perihelion of the orbit of a small body is at or near g_5, g_6, g_7 or g_8, then the body is said to be in the ν_5, ν_6, ν_7 or ν_8 resonance, respectively.

But the true condition for being in one of these resonances is if its associated resonance angle $\phi = \bar{\omega} - \bar{\omega}_p$ librates. Here, $\bar{\omega}$ and $\bar{\omega}_p$ are each the longitude of perihelion of the orbit of the small body and planet, respectively.

If the precession rate of the longitude of ascending node of the orbit of a small body is at or near s_5, s_6, s_7 or s_8, then the body is said to be in the ν_{15}, ν_{16}, ν_{17} or ν_{18} resonance, respectively (Williams 1969; Froeschle and Scholl 1989). Analogously, a small body is in one of these resonances if its associated resonance angle $\phi = \Omega - \Omega_p$ librates.

Table 4.3 Eigenfrequencies of the Jovian planets

Name	Value (arcsec/year)	Period (years)
g_5	$\frac{4.29591''}{\text{year}}$	3.0×10^5
g_6	$\frac{27.77406''}{\text{year}}$	4.7×10^4
g_7	$\frac{2.71931''}{\text{year}}$	4.8×10^5
g_8	$\frac{0.63332''}{\text{year}}$	2.0×10^6
s_5	$\frac{-25.73355''}{\text{year}}$	5.0×10^4
s_6	$\frac{-25.73355''}{\text{year}}$	5.0×10^4
s_7	$\frac{-2.90266''}{\text{year}}$	4.5×10^5
s_8	$\frac{-0.67752''}{\text{year}}$	1.9×10^6

Values are taken from Murray and Dermott (1999)

Example Shown in the figure is the longitude of ascending node of a planet vs. time. Which planet is this? If the precession rate of the longitude of ascending node of a test particle matched the precession rate of the longitude of ascending node of this planet, which resonance would the test particle be in?

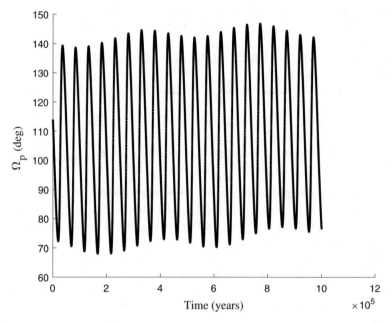

Solution From the figure, about 20 cycles are completed in a time of 10×10^5 years. This yields a period of

$\frac{10 \times 10^5}{20} = 5 \times 10^4$ years.

This matches the period of the longitude of ascending node of both Jupiter and Saturn, making the test particle in the ν_{15} and ν_{16} resonances.

Example Given a small body in an orbit about the Sun with a longitude of perihelion, $\bar{\omega}$. Given the longitude of perihelion of the orbit of Saturn is $\bar{\omega}_S$. If the angle given by the relation $\phi = \bar{\omega} - \bar{\omega}_S$ librates in time, then which secular resonance is this small body in?

Solution By definition, this is the ν_6 resonance.

A special case of a secular resonance occurs when the argument of perihelion of the orbit of a small body librates. This resonance is known as a **Kozai-Lidov resonance** (Kozai 1962; Sie et al. 2015; Shevchenko 2017).

In this type of resonance, the component of angular momentum of a small body parallel to the angular momentum of the perturbing body is conserved. This quantity, known as the Kozai integral, can be expressed as:

$$I_K = \cos i \sqrt{1 - e^2} \qquad (4.57)$$

where i and e are the inclination and eccentricity of the small body's orbit. As a consequence of this, any increase in e must be accompanied by a decrease in i and vice versa. These coupled oscillations between e and i are the hallmark of the Kozai-Lidov resonance.

Example Shown in the figure is the argument of perihelion of a test particle vs. time. During what approximate time does the test particle first enter the Kozai resonance?

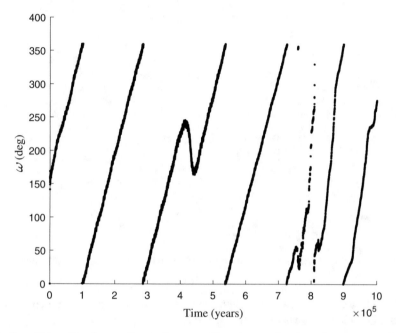

Solution From the figure, it can be seen that the angle abruptly changes from circulation to libration at 4×10^5 years.

4.3.3 Secondary Resonances

Secondary resonances occur when the libration frequency of the resonance angle of a mean motion resonance (the primary resonance) is commensurate with the circulation frequency of a nearby mean motion resonance (Kortenkamp et al. 2004). Secondary resonances lead to chaotic orbits and the disruption of the primary mean motion resonance (e.g., Malhotra and Dermott 1990).

As an example, in the top diagram of Fig. 4.13, the resonance angle of Neptune's 2:1 MMR with Uranus circulates in time with a period of 4300 years. In the bottom diagram of Fig. 4.13, the resonance angle of the 1:1 MMR of Uranus librates in time with a period of 5700 years for a test particle. Since $\frac{5700}{4300} \approx 1.33 \approx \frac{4}{3}$, this test particle is in the 1:1 MMR of Uranus and also in a 4 to 3 three-body secondary resonance.

Example Shown in the figure is a graph of the resonance angle of the 5:2 mean motion resonance of Saturn vs. time. The resonance is located at about 5.18 au. The output time is 50 years. What period should the libration of the resonance angle of a nearby mean motion resonance have in order to be in a 2:1 secondary resonance with the 5:2 mean motion resonance of Saturn?

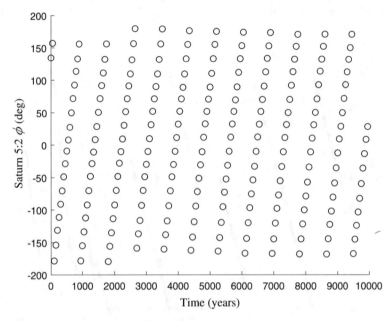

Solution It can be seen in the figure that the resonance angle completes eleven cycles in 9450 years. Therefore the period is

$\frac{9450}{11} = 859$ years.

Therefore, to be in a 2:1 secondary resonance with the 5:2 mean motion resonance of Saturn the libration period should be $2 \times 859 = 1718$ years.

4.4   Chaos and Stability

Chaoticity and stability are important characteristics of an orbit. Stability is measured by the dynamical lifetime of clones of an orbit. The **dynamical lifetime** is the amount of time a test particle remains part of a numerical integration before being removed due to some criterion that dictates its removal, such as a collision with a planet. The longer the average dynamical lifetime of clones, the more stable the orbit.

Chaos is often related to overlapping resonances in phase space. Resonances do not exist at only one point in phase space but instead take up a volume in six-dimensional phase space. This allows for the possibility of overlapping resonances. The orbits of small bodies in the vicinity of overlapping resonances can become highly chaotic.

Chaoticity and dynamical lifetime are separate quantities. An orbit can be both chaotic and have a short dynamical lifetime, or if an orbit is chaotic and has a relatively long dynamical lifetime, it is said to exhibit **stable chaos**.

Chaos is the study of extreme sensitivity to initial conditions (Gleik 1987). This means that because of chaos, tiny differences in the initial conditions between two different systems can result in very different final states over time.

For example, consider two massless test particles orbiting the Sun in orbits that differ only infinitesimally as shown in Fig. 4.14a. Test particles are placed into both orbits at positions which differ only minutely in phase space. In Fig. 4.14b the same two test particles are shown at a later time. The states of the two test particles have diverged.

Thus, there is potentially an initial difference between each position and velocity vector component between the two for a total of up to six component differences

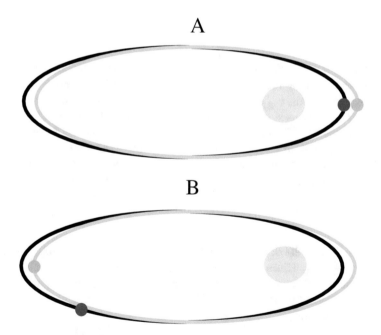

Fig. 4.14 In diagram (**a**), two orbits with the same eccentricity are shown, which differ in semimajor axis by an infinitesimally small amount. Test particles are placed into both orbits in positions which differ only minutely in phase space. In diagram (**b**), the same two test particles are shown at a later time. The distance between the two particles in phase space has greatly increased, showing that an initially small difference has resulted in a large change

(3 position and 3 velocity). These are: $\Delta x_0 = x_{20} - x_{10}$, $\Delta y_0 = y_{20} - y_{10}$, $\Delta z_0 = z_{20} - z_{10}$, $\Delta v_{xo} = v_{2x0} - v_{1x0}$, $\Delta v_{yo} = v_{2y0} - v_{1y0}$, $\Delta v_{zo} = v_{2z0} - v_{1z0}$.

If the orbit is chaotic, then these infinitesimal differences will grow exponentially in time. Thus, for any particular vector component initial difference, Δx_0, the difference Δx at a time t can be written in the form

$$\Delta x = \Delta x_0 e_{exp}^{\gamma t} \qquad (4.58)$$

Of the six orbital component differences, one of them will grow faster than the others and thus have the largest exponent, γ_{max}, of the six. This exponent is known as the **Lyapunov Characteristic Exponent** or Maximum Lyapunov Exponent and is given by:

$$\gamma_{max} = \lim_{t \to \infty} \frac{1}{t} \int_0^t \frac{\dot{\Delta x}(t')}{\Delta x(t')} dt' \qquad (4.59)$$

The associated **Lyapunov time**, t_{lyp}, is the time it takes the associated component difference to grow by a factor of e_{exp} (the natural number 2.718281828...) and is given by

$$t_{lyp} = \frac{1}{\gamma_{max}} \qquad (4.60)$$

The value of t_{lyp} is dependent on the integration time (Whipple 1995). Orbits with shorter Lyapunov times are considered more chaotic than those with longer Lyapunov times. Thus, t_{lyp} can be used to measure the chaoticity of an orbit.

Another quantity used to measure chaos related to the Lyapunov Characteristic Exponent is the **MEGNO parameter**, Y, where MEGNO stands for mean exponential growth of nearby orbits (Cincotta and Simó 2000; Cincotta et al. 2003). The time-averaged MEGNO parameter, $\langle Y \rangle$, is a dimensionless quantity directly proportional to the Lyapunov Characteristic Exponent and time, t, via

$$\langle Y \rangle = t \frac{\gamma_{max}}{2} \qquad (4.61)$$

(Cincotta and Simó 2000; Goździewski et al. 2001; Cincotta et al. 2003; Giordano and Cincotta 2004; Hinse et al. 2010). In the limit $t \to \infty$ the value of $\langle Y \rangle$ asymptotically approaches 2.0 for quasi-periodic orbits and rapidly diverges far from 2.0 for chaotic orbits. The MEGNO parameter can be calculated by solving the following integral

$$Y = \frac{2}{t} \int_0^t \frac{\dot{\Delta x}(t')}{\Delta x(t')} t' dt' \qquad (4.62)$$

The time-averaged MEGNO parameter is given by

$$\langle Y \rangle = \frac{1}{t} \int_0^t Y(t')dt' \qquad (4.63)$$

MEGNO has been used to study various objects including galaxies (Cincotta and Simó 2000; Cincotta et al. 2003), irregular Jovian moons (Hinse et al. 2010) and exoplanets (Goździewski et al. 2001). A sample MEGNO map is shown in Fig. 4.15.

Example Identify the block of quasi-periodic orbits in Fig. 4.15 that has the smallest eccentricity values.

Solution In the figure, it can be seen that the color black represents orbits for which the MEGNO parameter is 2, so these are the quasi-periodic orbits. There are three different blocks of quasi-periodic orbits. Of these, the block with the smallest eccentricity values can be found between 13.93 au $\leq a \leq$ 13.95 au and $0 \leq e \leq 0.0013$.

The stability of an orbit can also be measured by finding the **half-life** of the orbit. The half-life is the time it takes the initial number of clones in a simulation to be reduced by one-half. This can be found using the following method

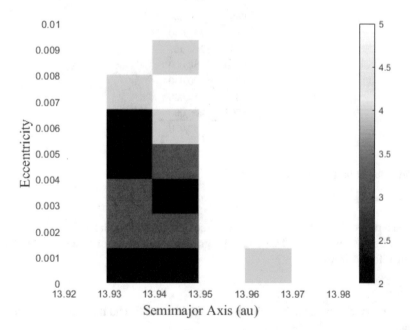

Fig. 4.15 A Megno map in the shown region of $a - e$ space for test particles with initial orbital parameters $i = 6.947°$, $\Omega = 209.216°$, $\omega = 339.537°$, and $M = 145.978°$. Orbits with $\langle Y \rangle \geq 5$ (shown in white) are highly chaotic and orbits with $\langle Y \rangle = 2$ are quasi-periodic

1. Create clones of the orbit.
2. Create criteria by which clones will be removed from a numerical simulation. These may be by collision with a planet or the Sun, by obtaining a certain heliocentric distance, by its orbit obtaining an eccentricity ≥ 1 or other criteria.
3. Integrate the clones backwards or forwards in time using a numerical integrator and remove clones according to the criteria.
4. Fit the number of remaining clones as a function of time to the radioactive decay equation shown in Eq. (4.64) and determine the half-life of the decay.

$$N = N_o e_{exp}^{\frac{ln(0.5)}{\tau}t} \tag{4.64}$$

Here, τ is the half-life, N_o the initial number of clones and N the number of remaining clones at a certain time, t.

Example A small body is cloned and numerically integrated in time. Shown is the number of remaining clones, N at each time, t. What is the half-life of this body's orbit?

N	t (years)
10,000	0
9830	50,000
9660	100,000
9500	150,000
9330	200,000
9170	250,000
9000	300,000
8860	350,000
8705	400,000
8565	450,000

Solution Equation (4.64) can be rewritten as

$\ln(\frac{N}{N_o}) = \frac{\ln(0.5)}{\tau}t$

Then a plot of $\ln(\frac{N}{N_o})$ vs. t is linear with the best-fit slope equal to $\frac{\ln(0.5)}{\tau}$. The slope of the best-fit line can be found using a graphing calculator or Excel. The half-life is then found from

$\tau = \frac{\ln(0.5)}{slope}$

Using this technique, the best-fit slope is -3.46×10^{-7}, and the half-life is 2×10^6 years.

The longer the half-life, the more stable the orbit. Note that the forward and backward half-lives need not equal each other (Horner et al. 2004). This means that an orbit may be more stable in one time direction than another.

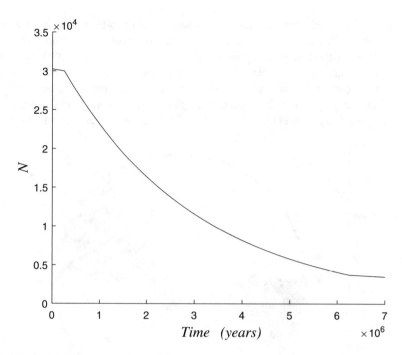

Fig. 4.16 The number of particles vs. time

Figure 4.16 shows an example of the exponential decay of clones of an orbit as a function of time. Note the backwards 's' shape, which is common in graphs such as these. The flat portion of the graph starting at a time of zero is due to the time it takes for the clone swarm to disperse. Once dispersal has been achieved, exponential decay begins at a time of 250,000 years. The flat portion at the right end of the graph is due to the remaining orbits being so stable that the decay is no longer exponential.

Exercises

1. Given the planet Uranus has a mass of 14.5 Earth masses in an orbit with a semimajor axis of 19.2 au about the Sun, find the Hill radius of Uranus with respect to the Sun, if the Sun has a mass of 332,625 Earth masses.
2. Given point P on an elliptical heliocentric orbit. The distance between point P and the Sun is d_1, and the distance between point P and the empty focus is d_2. Since the orbit is an ellipse, it must be true that $d_1 + d_2$ = constant. Using the point of perihelion, prove that this constant = 2 × the semimajor axis of the orbit.

3. In the following figure the planet Mercury orbits the Sun in the elliptical orbit shown. A line drawn from the planet to the Sun sweeps out an area, A_1, in a time of 11 days as Mercury moves from point A to point B. When Mercury moves from point C to point D, a line drawn from the planet to the empty focus sweeps out an area also in 11 days. A student claims that the area swept out when Mercury moves from point C to point D is also A_1. Form an argument using Kepler's second law refuting or defending this student's statement.

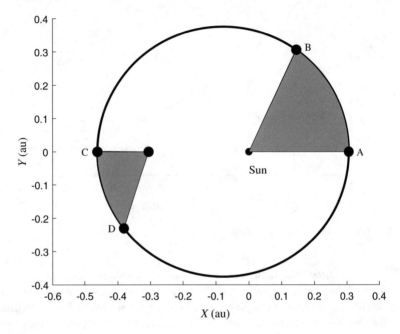

4. Earth is in an orbit with a semimajor axis of 1 astronomical unit and takes 1 year to orbit the Sun. A student claims that a small body orbiting the Sun in an elliptical orbit with a semimajor axis of 2 astronomical units will take 2 years to orbit the Sun. Using Kepler's third Law, form an argument defending or refuting this statement.

5. Shown in the figure are two heliocentric orbits of small bodies with the Sun shown for each orbit. Find the ratio of the orbital period of the body in Diagram 2 to the orbital period of the body in Diagram 1.

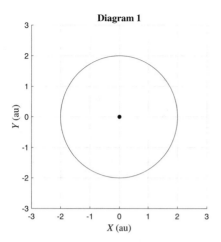

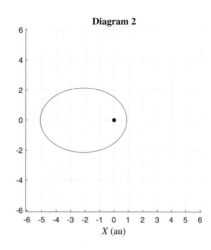

6. Given a uniform spherical planet of radius R_p and a small moon of mass m that orbits it at a distance d from the planet's center of mass. Given point A at a distance of $0.5R_p$ from the center of the planet along a planet-moon radial line as shown in the figure. Derive an equation for the strength of the tidal force per unit mass (the tidal acceleration) at point A.

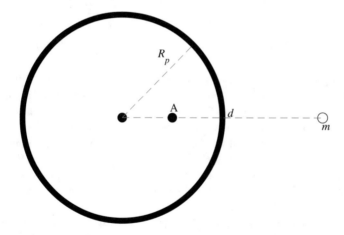

7. Shown in the figure are four heliocentric orbits of four small bodies. The Sun is located at the black dot in each diagram. For each quantity shown, rank these diagrams in order from left to right from smallest to largest quantity value, or if the quantity is the same for all orbits, state that it is so.

   (a) Eccentricity
   (b) Speed of orbiting small body at perihelion
   (c) Orbital period of orbiting small body

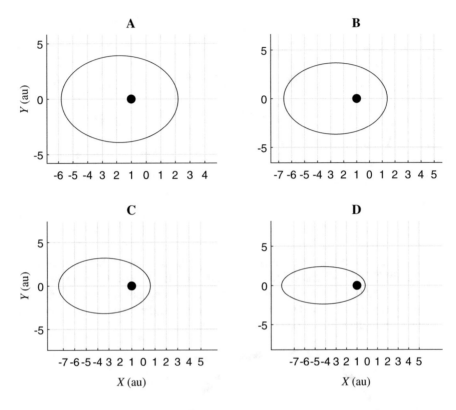

8. For elliptical, hyperbolic and parabolic orbits, state the range of possible v_∞ values for each orbit from among $v_\infty = 0$, $v_\infty > 0$ and v_∞ is undefined.

9. Find the order and the location of the Neptune 3:2, Saturn 5:9 and Jupiter 1:3 mean motion resonances using the planetary data from Table 4.1.

10. Find the distance between the 1:2 mean motion resonance of Uranus and the semimajor axis of the orbit of Neptune using the planetary data from Table 4.1.

11. Using the planetary data from Table 4.1, round off the ratio of the orbital periods of Uranus to Saturn to the nearest integer to find what mean motion resonance of Uranus these planets are nearly in.

12. The orbit of comet Encke has an eccentricity of $e = 0.848$, a semimajor axis of $a = 2.22$ au and an inclination $i = 11.78°$. Given the inclination and semimajor axis of Jupiter's orbit are $1.3°$ and 5.2 au respectively, find the Tisserand parameter of comet Encke with respect to Jupiter.

13. Using the lowest possible integers, which planet has a mean motion resonance located at 15.85 au? Define the resonance using two small integers.

14. Shown in the figure is a scatter plot of the longitude of perihelion of a planet vs. time. Identify the planet. If the precession rate of the longitude of perihelion of a test particle matches the precession rate of the longitude of perihelion of this planet, what resonance is the test particle in?

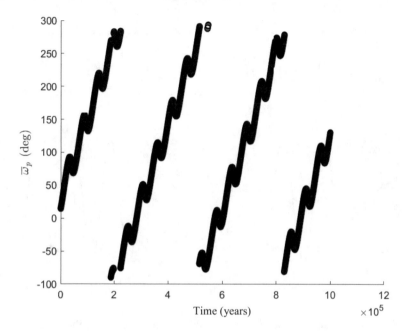

15. Given four diagrams of librating resonance angles. Unsubscripted quantities refer to quantities of a test particle and subscripted quantities refer to those of a planet with subscripts S, U and N referring to Saturn, Uranus and Neptune, respectively. Name the resonance associated with each diagram from among Kozai-Lidov, ν_5, ν_6, ν_7, ν_8, ν_{15}, ν_{16}, ν_{17} and ν_{18}.

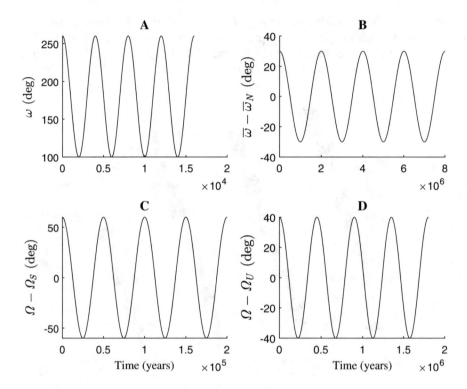

16. In the figure are four diagrams showing the libration of the resonance angle
    for a test particle in the 1:1 mean motion resonance of Uranus. Given that the
    resonance angle for the nearby 2:1 mean motion resonance between Neptune
    and Uranus circulates with a period of 4300 years, identify the three-body
    secondary resonance for each diagram using a ratio of small integers.

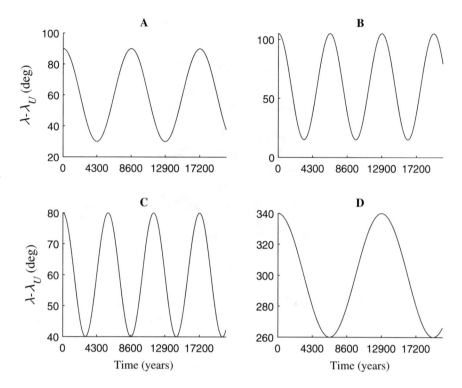

17. Suppose a new planet has been discovered in our solar system with a semimajor axis of 400 au. In the figure, a MEGNO map is shown for a region of $a - e$ space around the orbit of this new planet. This map was created using 120,000 test particles spread out evenly in a 300×400 grid in $a - e$ space so that each test particle had a unique combination of a and e. Initially, test particles were in the plane of the orbit of the planet with $M = 60°$. Initial values of other orbital parameters for test particles were set to zero. What is the width in au of the Gaussian shape in the figure shown in black for circular orbits? What is the maximum value of the eccentricity of quasi-periodic orbits?

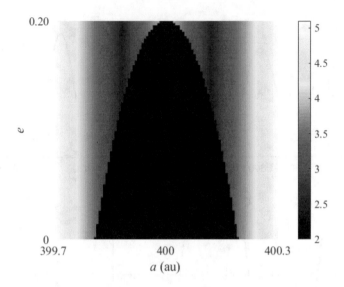

18. A small body is cloned and numerically integrated in time. Shown is the number of remaining clones, N at each time, t. What is the half-life of this body's orbit?

N	t (years)
10,000	0
9880	50,000
9772	100,000
9653	150,000
9548	200,000
9445	250,000
9330	300,000
9223	350,000
9122	400,000
9000	450,000

Chapter 5
Populations of Small Solar System Bodies

5.1 Asteroids

Asteroids are objects that are mainly composed of rocky and metallic material, are generally accepted to be debris left over from the formation of the solar system. The exact composition varies with the asteroid, but in addition to rocky material, asteroids may contain metals and volatiles including organic material (Burbine 2017). A **volatile** is a substance with a significantly lower boiling point than rocky material. A substance with a higher boiling point is called a **refractory**.

Most asteroids orbit the Sun between the orbits of Mars and Jupiter. These asteroids with semimajor axes between 2 au and 4.28 au are known as Main-Belt Asteroids or MBAs, and the region in which they orbit is called the **Main Asteroid Belt**. The asteroids in this belt never accreted into a planet because of perturbations from Jupiter (Petit et al. 2001).

Over 500,000 objects are listed in the Minor Planet Center database as existing in this region.[1] Daniel Kirkwood discovered regions in the Main Asteroid Belt where there are relatively lower populations of asteroids compared to nearby orbits, and that these gaps were related to mean motion orbital resonances with Jupiter (Kirkwood 1867). These regions are now known as **Kirkwood gaps** (Ryden 2016). Figure 5.1 shows a histogram of small bodies in the inner solar system. The gaps in the Main Asteroid Belt at the location of mean motion resonances can clearly be seen.

Other dynamical classes of asteroids exist. Asteroids that are locked in a 1:1 mean motion orbital resonance with a planet and have resonance angles that librate about the L_4 or L_5 Lagrange point in tadpole orbits are known as **Trojan asteroids**. Trojans have been discovered for Jupiter, Uranus, Neptune, Mars, Earth and Venus

---

[1] https://www.minorplanetcenter.net/iau/MPCORB.html (accessed Sep. 24, 2017).

© The Author(s), under exclusive license to Springer Nature Switzerland AG 2019
J. Wood, *The Dynamics of Small Solar System Bodies*, SpringerBriefs
in Astronomy, https://doi.org/10.1007/978-3-030-28109-0_5

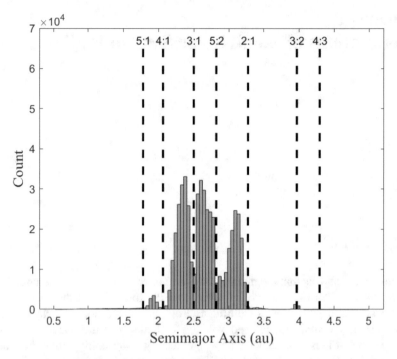

Fig. 5.1 A histogram of small bodies in the inner solar system with a bin size of 0.05 au. The large concentration of asteroids is the Main Asteroid Belt. Also shown are the locations of a few of Jupiter's interior mean motion resonances. Note the dearth of asteroids at these locations

(Marzari and Scholl 2002; Scholl et al. 2005; Connors et al. 2011; Lykawka et al. 2011; Aron 2013; de la Fuente Marcos and de la Fuente Marcos 2014).

While Jupiter boasts a Trojan population that exceeds 6000, by contrast, no Saturn Trojans have been discovered. This is likely because the Trojan regions of Jupiter are dynamically stable on Gyr timescales, while those of Saturn are mostly not, except for a few small niches. Saturn may once have had a large Trojan population, but even if it did, today this population would have been greatly depleted due to the overlap of the Trojan region with the Jupiter 2:5 mean motion resonance and secular resonances (Marzari and Scholl 2000; Marzari et al. 2002; Nesvorný and Dones 2002).

Nevertheless, the instability of the Trojan region does not exclude the possibility that Saturn has captured small bodies into temporary Trojan orbits (Horner and Wyn Evans 2006) and future surveys such as the LSST may indeed discover such objects. Jupiter has by far the largest number of discovered Trojans. The number of Trojans for each planet is shown in Table 5.1.[2]

---

[2]https://www.minorplanetcenter.net/iau/lists/Trojans.html (accessed Dec. 10, 2017).

Table 5.1 The current number of Trojans by planet according to the Minor Planet Center

Planet	Number of Trojans
Earth	1
Mars	9
Jupiter	6701
Uranus	1
Neptune	17

It is believed that the Trojan asteroids of Jupiter and Neptune were captured into their Trojan regions as they migrated during the formation of the solar system. This explains the high inclinations of some of the orbits of Trojans of Jupiter and Neptune (Gomes et al. 2005).

Asteroids locked in a 1:1 mean motion orbital resonance with a planet are not necessarily Trojans. If the asteroid librates about the L_1 or L_2 Lagrange point and is not a satellite, then the small body is said to be a **quasi-satellite**, and if it librates about the L_3 Lagrange point, it is considered to be in a horseshoe orbit. All small bodies in a 1:1 mean motion orbital resonance with a planet are called **co-orbitals**.

Some asteroids come dangerously near to or even cross Earth's orbit. These asteroids are known as Near Earth Objects or NEOs. The term NEAs is also used for Near Earth Asteroids. There are three main subtypes of NEAs: Amor, Apollo, and Aten, which are defined using their semimajor axes and perihelion distances, q, as follows:

- Amors—have semimajor axes greater than 1 au and $1.017\,au < q < 1.3\,au$
- Apollos—have semimajor axes greater than 1 au and $q < 1.017$ au, which places them in Earth-crossing orbits with semimajor axes beyond Earth's
- Atens—have semimajor axes less than 1 au

(Norton and Chitwood 2008). The Apollos are Earth crossing, but the Amors are not. Atens can be either Earth crossing or not. Atens, which are not Earth crossing are referred to as Atiras. Currently 19 Atiras are known.[3] As of this writing, the Minor Planet Center classifies over 17,000 objects as NEAs.[4] The breakdown by subtype is shown in Table 5.2. These types of asteroids may exist in part due to perturbations from giant planets, which are known to perturb the orbits of asteroids into planet-crossing orbits. Specifically, the ν_6 secular resonance of Saturn and the 3:1 MMR of Jupiter are believed to be responsible for the transport of asteroid fragments to Earth-crossing orbits (e.g. Wisdom 1982, 1983; Froeschle and Scholl 1986).

---

[3] https://cneos.jpl.nasa.gov/about/neo_groups.html (accessed January 1, 2018).

[4] https://www.minorplanetcenter.net/iau/lists/MPLists.html (accessed Dec. 10, 2017).

Table 5.2 The number of
NEAs listed by subtype
according to the Minor Planet
Center

Type of asteroid	Number
Aten	1285
Apollo	8582
Amor	7393

5.2 Comets

Comets are bodies of ice and rock that orbit the Sun in much more eccentric orbits than those of the planets. What differentiates comets from asteroids is their large amount of ice. In general (but not always), at large heliocentric distances, comets exist as single, solid nuclei of ice and rock orbiting the Sun (Levison and Duncan 1997).

But at closer distances to the Sun, ices begin to sublimate due to the Sun's energy causing the comet to have its own atmosphere. To sublimate or undergo **sublimation** means to transform directly from a solid into a gas. At this point the comet is said to be "**active**". The comet's atmosphere created by the activity is called the **coma** and appears as a fuzzy ball around the solid part of the comet called the **nucleus**. This type of activity normally occurs for comets within a heliocentric distance of about 3 au and is caused by water-ice sublimation (Duncan et al. 1988; Quinn et al. 1990; Levison and Duncan 1997; Meech and Svoren 2004; Emel'yanenko et al. 2013).

Heliocentric distance is typically a strong driver of cometary activity for bodies on highly eccentric orbits as the body reaches maximum activity at perihelion and minimum activity at aphelion. At heliocentric distances beyond 3 au however, the temperature is too low for water-ice to sublimate, and so other volatile species drive the activity. Typically, this activity is unrelated to heliocentric distance (Jewitt 2009; Womack et al. 2017).

For example, bodies like 95/P Chiron and comet 29P/Schwassman-Wachmann 1 display sporadic outbursts of activity that occur well beyond perihelion and seemingly have nothing to do with heliocentric distance (Hartmann et al. 1988; Lazzaro et al. 1997; Duffard et al. 2002). One explanation for this type of activity is escaping gasses from the interior of the nucleus (e.g. Womack et al. 2017).

As the comet continues to draw closer to the Sun, two tails develop—the **ion tail** and the **dust tail**. Each tail has a general direction of away from the Sun. The tails develop by two different mechanisms. The dust tail develops due to radiation pressure from the Sun. As particles sublimate and leave the nucleus, radiation pressure pushes the particles outward as they continue to orbit the Sun. The result is a curl-shaped tail that begins radially outward from the comet's nucleus and then curls behind the comet in its orbit.

The ion tail develops due to fast moving charged particles from the Sun whose magnetic field rips particles right off the nucleus, trapping them and carrying them along. With each passage around the Sun, a comet loses some of its own mass, which is shed in the form of particles that spread throughout the comet's orbit over time (Festou et al. 2004).

Traditionally, comets have been categorized as either short period or long period. Long-period comets revolve around the Sun with orbital periods greater than 200 years, and short-period comets revolve around the Sun with orbital periods less than 200 years (Levison 1996).

One subclass of short period comets is the Jupiter-Family comets. These can be loosely defined as comets whose dynamics are controlled by Jupiter and have orbital periods less than 20 years.

The Minor Planet Center assigns comets a provisional designation using a prefix letter, a forward slash and an identifier. As stated in the Minor Planet Circular 23803-4: "For comets, the acceptable prefixes are P/ for a periodic comet (defined to have a revolution period of less than 200 years or confirmed observations at more than one perihelion passage) and C/ for a comet that is not periodic (in this sense), with the addition of X/ for a comet for which a meaningful orbit cannot be computed and D/ for a periodic comet that no longer exists or is deemed to have disappeared."

For example, the periodic comets LINEAR and Halley have provisional designations P/2005 YQ127 and P/1682 Q1 respectively. When a comet is given a more proper name, the format is a sequential number, a prefix and the comet name. For example, 174P Echeclus and 75D Kohoutek.

It is generally accepted that originally many (but not all!) of the icy bodies known as comets originated from somewhere beyond Neptune in orbits that did not bring them into the inner solar system.

Today, this region beyond Neptune has several major subpopulations in which these bodies could have originally abided, including the **Edgeworth-Kuiper Belt** (Levison and Duncan 1997), the **Scattered Disk** (Volk and Malhotra 2008) and the **Oort cloud** (Emel'yanenko et al. 2005).

The Edgeworth-Kuiper Belt is a region between the semimajor axis of Neptune's orbit and 55 au. Small bodies in orbits with semimajor axes in this region are called **Kuiper Belt Objects** (or **KBOs**) (Lykawka and Mukai 2007). Medium-sized Kuiper Belt Objects are on the order of 10^2 km in diameter, and the largest are on the order of 10^3 km in diameter (Murray-Clay and Schlichting 2011). Pluto is one of these.

The Scattered Disk is a region overlapping the outer Edgeworth-Kuiper Belt and extending out to approximately 1000 au (Tiscareno and Malhotra 2003; Lykawka and Mukai 2007). But the boundaries are poorly defined. The Minor Planet Center also considers a few objects with semimajor axes beyond 1000 au to be in this region.[5] Small bodies in orbits with semimajor axes in this region are known as **Scattered Disk Objects** or **SDOs**. At this time, the Minor Planet Center recognizes over 540 SDOs.[6]

The Oort Cloud is a spherical cloud of comets surrounding the Sun (Oort 1950). Objects in this cloud are known as **Oort Cloud Objects** or **OCOs**. The exact boundaries of the Oort Cloud remain uncertain, but the cloud is believed to extend from 1000 au out to at least 20,000 au (Hills 1981) and may extend out to 200,000 au

---

[5]https://minorplanetcenter.net/iau/lists/Centaurs.html.

[6]https://www.minorplanetcenter.net/iau/lists/t_centaurs.html (accessed Dec. 10, 2017).

(Duncan et al. 1987; Dones et al. 2015). Long-period comets are believed to have originated in the Oort Cloud (e.g. Oort 1950; Fouchard et al. 2014).

OCOs, KBOs, SDOs and other small bodies in orbits that have semimajor axes beyond that of Neptune are classified as **Trans-Neptunian Objects** (or **TNOs**) by the Minor Planet Center of which 1900 are known.[7] So it can be said that some comets can also be classified as TNOs. Neptune trojans are a special case and are not considered to be TNOs.

Since comets burn out on time scales much smaller than the age of the solar system, the comets seen today could have originally been in the Trans-Neptunian region before being sent to the inner solar system by gravitational perturbations from planets (Duncan et al. 1988; Levison and Duncan 1997; Volk and Malhotra 2008), a passing star (Oort 1950; Hills 1981) or galactic tides (Heisler and Tremaine 1986; Duncan et al. 1987).

Thus, the specific classification of any small solar system body such as a comet may be ephemeral as small bodies may transition back and forth between different classes of objects during their lifetime (Tiscareno and Malhotra 2003; Horner et al. 2004).

5.3   Centaurs

In 1977, a small body was discovered orbiting the Sun between Saturn and Uranus. It was named 2060 Chiron (Kowal et al. 1979). Today we know that Chiron's orbit has a semimajor axis of about 13.65 au and that Chiron itself has a radius of at least 71 km (Groussin et al. 2004).

Its discovery was surprising because at that time no objects of this size were known to have orbits like this between giant planets. Chiron's discovery was followed in 1992 by the discovery of 5145 Pholus orbiting between Uranus and Neptune.

As time went on, more of these curious objects were found: 7066 Nessus in 1993, 8405 Asbolus in 1995 and 10199 Chariklo in 1997. It was soon realized that these bodies represented a new class of objects which orbit the Sun between Jupiter and Neptune. These bodies were hard to classify as either a traditional asteroid or comet due to their orbits and icy composition.

Such objects were named **Centaurs**, as they were thought to have properties of both comets and asteroids in the same way that mythological Centaurs have characteristics of both humans and horses.

The Minor Planet Center defines a Centaur as a small body in an orbit with a semimajor axis between those of the orbits of Jupiter and Neptune and a perihelion

---

[7]https://minorplanetcenter.net/iau/lists/t_tnos.html (accessed Dec. 10, 2017).

distance greater than the semimajor axis of Jupiter's orbit.[8] But Centaurs are poorly
defined. Other definitions include

- Comets in orbits which have Tisserand parameters with respect to Jupiter, T_J,
  which obey the relation $T_J > 3$ and have semimajor axes greater than that of
  Jupiter's (Levison 1996; Duncan et al. 2004)
- SSSBs with 5 au $< q <$ 28 au and $a <$ 1000 au (Emel'yanenko et al. 2013).

This work adopts the definition of Centaur used by the Minor Planet Center.
Applying this definition to a database of small bodies obtained from the Minor
Planet Center shows that the number of objects classified as Centaurs is >220,[9]
however, the actual population is believed to be much higher. Horner et al. (2004)
estimate the real population of Centaurs with diameters larger than 1 km to be
44,300.

It is believed that Centaurs are a relatively short-lived transitional class of objects
between KBOs and Jupiter-Family comets (Tiscareno and Malhotra 2003). Small
objects survive in the Centaur region typically on the order of 10 Myr (Levison and
Duncan 1994; Tiscareno and Malhotra 2003; Dones et al. 1996; Horner et al. 2004).

Since the solar system is about 4.6 billion years old, the Centaurs we see today
must have originated in another part of the solar system. Thus, for their population
to be maintained, they must have a replenishing source. Numerical studies show that
the orbits of KBOs and SDOs can be perturbed by Neptune in such a way that their
perihelia decrease until the objects transition into Centaurs (Duncan et al. 1988;
Levison and Duncan 1997; Volk and Malhotra 2008).

SDOs in particular are a likely source of Centaurs (Di Sisto and Brunini 2007).
Other source populations include the Oort Cloud (Emel'yanenko et al. 2005; Brasser
et al. 2012; de la Fuente Marcos and de la Fuente Marcos 2014; Fouchard et al.
2014), Trojan population of Jupiter (Horner et al. 2004), and Trojan population of
Neptune (Horner and Lykawka 2010).

The lifetimes of Centaurs are punctuated by close encounters with the giant
planets. These close encounters along with other orbital perturbations cause the
osculating orbital parameters to change over time. Thus, the orbits of Centaurs are
in a constant state of flux. The dynamical evolution of the orbits of Centaurs are
complex, and the entire Centaur region is highly chaotic. Sometimes Centaurs can
maintain a near constant perihelion while their semimajor axes and eccentricities
change (Horner et al. 2003).

Numerical studies show that Centaurs exist in one of two dynamical classes
(Tiscareno and Malhotra 2003; Bailey and Malhotra 2009). One type consists
of those Centaurs whose proper semimajor axis changes in time according to
a power law. These are referred to as **random-walk Centaurs**. The other type
consists of Centaurs that abruptly jump between mean motion orbital resonances
of the four Jovian planets. Resonance sticking dominates the dynamics of these

---

[8]http://www.minorplanetcenter.net/iau/lists/Unusual.html (accessed Dec. 10, 2017).

[9]https://www.minorplanetcenter.net/iau/lists/t_centaurs.html (accessed Dec. 10, 2017).

Centaurs throughout their lifetimes. These are called **resonance hopping Centaurs**. Numerical studies show that on average, resonance hopping Centaurs have longer dynamical lifetimes than random-walk Centaurs. Random-walk Centaurs are more likely to evolve into Jupiter-Family comets than resonance hopping Centaurs (Di Sisto and Brunini 2007; Bailey and Malhotra 2009) (Fig. 5.2).

Orbital perturbations can cause Centaurs to jump back and forth between different small body populations. Centaurs can transition into Jupiter-Family comets (Levison 1996; Horner et al. 2004), then back into Centaurs, then back into Jupiter-Family comets again multiple times. Centaurs can collide with the Sun or a planet, be ejected from the solar system entirely or even evolve into NEOs (Horner et al. 2004). The latter is of particular concern, given that Centaurs are made mostly of volatile material and can be much larger ($\sim 10^2$ km) than the sizes of typical comet nuclei (~ 10 km).

As a comparison, comet C/1995 O1, Hale-Bopp, one of the largest comets known, has a nucleus with a diameter of at most 35 km (Weaver and Lamy 1997), but the largest Centaur known, Chariklo, has a diameter of about 250 km (Araujo et al. 2016).

5.4   The Taxonomy of Comets and Comet-Like Bodies

The very basic taxonomical scheme that classified comets as short-period or long-period was based on the longest time over which calculated orbital parameters for comets were considered reliable which then was about 200 years. Today, this scheme has become obsolete (Levison 1996). Comet taxonomy is fluid, and more than one taxonomical scheme has been proposed.

The taxonomical scheme of Levison (1996) separates comets into two broad categories based on the value of their Tisserand parameter with respect to Jupiter. Nearly Isotropic Comets have $T_J < 2$ and Ecliptic Comets have $T_J > 2$.

Ecliptic Comets are further subclassified as Jupiter-Family, Chiron-type or Encke-type. Jupiter-Family comets have $2 < T_J < 3$. Encke-type comets have $T_J > 3$ with semimajor axes less than that of Jupiter's orbit and Chiron type comets have $T_J > 3$ with semimajor axes greater than that of Jupiter's orbit (Levison et al. 2006).

Nearly Isotropic Comets are split into two classes: New and Returning. Comets with $a > 10,000$ au are classified as New and comets with $a < 10,000$ au are classified as Returning. Returning comets are further split in two classes of their own: those with $a < 40$ au are termed Halley-type and those with $a > 40$ au are termed External (Levison 1996). The entire scheme is shown in Fig. 5.3.

Bodies like Chiron-type comets or Centaurs are not true comets, in the sense that they do not periodically grow tails and a coma. However, these bodies may be composed of the same ices found in comets and may display outbursts of cometary activity on occasion (Jewitt 2009). 2060 Chiron has been known to display cometary activity unrelated to its heliocentric distance beyond a heliocentric distance of 3 au (Bus et al. 1991).

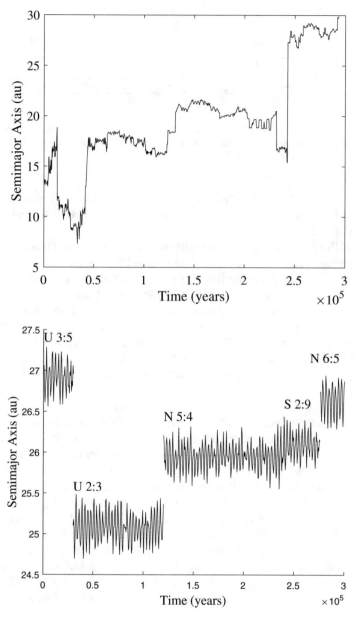

Fig. 5.2 Top—an example of a random-walk Centaur. Bottom—an example of a resonance hopping Centaur. Resonances visited are shown. N—Neptune, U—Uranus, and S—Saturn

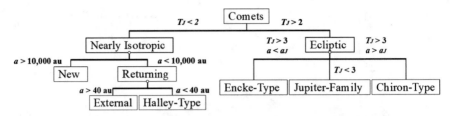

Fig. 5.3 The comet taxonomical scheme of Levison (1996). T_J is the Tisserand parameter with respect to Jupiter and a_J is the semimajor axis of the orbit of Jupiter (Image created by author with permission but based on an image in Levison 1996)

The taxonomical classification scheme of Horner et al. (2003) (henceforth known as the **HEBA scheme**) can be used to classify comet-like bodies such as these, as well as traditional comets.

In the HEBA scheme, comet-like bodies are placed into a major class and a minor class. The major class is based on the planet controlling the dynamics of the body at perihelion and the planet controlling the dynamics of the body at aphelion using a single letter abbreviation for each planet—J for Jupiter, S for Saturn, U for Uranus and N for Neptune. Thus, the major class may be shown using up to two letters.

The first letter of the major class is the starting letter of the planet controlling the dynamics of the body at perihelion, and the second that of the planet controlling the dynamics of the body at aphelion. For example, if Saturn controls the dynamics of the body at perihelion and Neptune at aphelion, then the body is said to be in the SN class.

If the body is controlled by the same planet at both perihelion and aphelion, then only the one letter of the controlling planet is used. Table 5.3 shows the letters used in the HEBA classes and the ranges over which each planet controls the dynamics of a small body at perihelion and aphelion based on Horner et al. (2003).

Sometimes the aphelion of the body is so large that no planet controls its dynamics at aphelion. If no planet controls the dynamics of the body at aphelion, then the letters E, T or EK are used depending on values of q and Q.

Once the major class of the body has been determined, the minor class is found using the Tisserand parameter, T_p, with respect to the planet controlling the dynamics of the small body at perihelion. Four unique ranges of T_p each define a possible value for the minor class, and each value is signified using a subscripted Roman numeral in the range I–IV written after the major class. For example, a small body in the SN major class may be completely categorized as SN_I, SN_{II}, SN_{III} or SN_{IV}.

If there is no controlling planet at aphelion, then if the perihelion lies in the range $4\,\mathrm{au} \leq q \leq 33.5\,\mathrm{au}$, and the aphelion lies in the range $Q \leq 60\,\mathrm{au}$, the letter E is used for Edgeworth-Kuiper Belt for the second letter of the major class.

If the perihelion lies in the range $4\,\mathrm{au} \leq q \leq 33.5\,\mathrm{au}$, and the aphelion lies in the range $Q > 60\,\mathrm{au}$, the letter T is used for Trans-Edgeworth-Kuiper Belt for the second letter of the major class.

Table 5.3 The letters used in the major HEBA classes for the first letter, second letter and stand alone classes based on the values of q and Q

Letters	Perihelion (au)	Aphelion (au)
First and second letters		
J	4–6.6	4–6.6
S	6.6–12	6.6–12
U	12–22.5	12–22.5
N	22.5–33.5	22.5–33.5
Second letter only		
E	$4 \leq q \leq 33.5$	$Q \leq 60$
T	$4 \leq q \leq 33.5$	$Q > 60$
Stand alone classes		
EK	$q > 33.5$	$Q \leq 60$
T	$q > 33.5$	$Q > 60$

Based on Horner et al. (2003)

Table 5.4 The HEBA minor classes based on the Tisserand parameter with respect to the planet controlling the dynamics of the small body at perihelion

Minor class	Tisserand parameter
I	$T_p \leq 2$
II	$2 < T_p \leq 2.5$
III	$2.5 < T_p \leq 2.8$
IV	$T_p > 2.8$

Based on Horner et al. (2003)

Table 5.5 The four comet classes in the HEBA taxonomy

Class	Perihelion (au)	Aphelion (au)
E	$q < 4$	$Q < 4$
SP	$q < 4$	$4 \leq Q \leq 35$
I	$q < 4$	$35 < Q \leq 1000$
L	$q < 4$	$Q > 1000$

E Encke-type comet, *SP* short-period comet, *I* intermediate-period comet and *L* long-period comet. Based on Horner et al. (2003)

There are two special cases where no planet controls at perihelion or aphelion. Objects with $q > 33.5$ au and $Q > 60$ au are said to be in the T class, and objects with $q > 33.5$ au and $Q \leq 60$ au are said to be in the EK class for Edgeworth-Kuiper Belt.

The four minor classes are given in Table 5.4.

For example, using Table 5.4, if Saturn controls the dynamics of a small body at perihelion, Uranus controls the dynamics at aphelion, and the Tisserand parameter with respect to Saturn is 2.6, then the object is classified as SU_{III}. As another example, if Saturn controls the dynamics at perihelion, $Q = 35$ au and $T_p = 2.85$ then the object is classified as SE_{IV}.

Four comet classes cover all bodies that have perihelia $q < 4$ au and are defined in Table 5.5. The four comet classes in the HEBA taxonomical scheme are

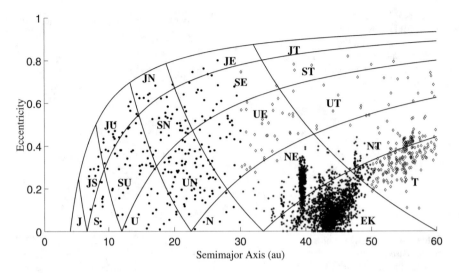

Fig. 5.4 The HEBA scheme. Filled circles are Centaurs, asterisks are Kuiper Belt Objects and diamonds are scattered disk objects

E = Encke-type comet for $Q < 4$ au, SP = short-period comet for $4 \leq Q \leq 35$ au, I = intermediate-period comet for $35 < Q \leq 1000$ au and L = long-period comet for $Q > 1000$ au.

Figure 5.4 shows the currently known Centaurs, SDOs and Kuiper Belt Objects according to the Minor Planet Center in the HEBA scheme.

Example Shown are the inclinations, semimajor axes and eccentricities of some selected small bodies. Complete this table by finding the major and minor HEBA classes of each body. If the object has no minor class, then leave it blank.

Designation	i (°)	e	a (au)	HEBA major	HEBA minor
5145 Pholus	24.6	0.570	20.393		
7066 Nessus	15.7	0.519	24.641		
8405 Asbolus	17.6	0.622	17.974		
10370 Hylonome	4.1	0.247	25.199		

Solution First calculate the perihelion and aphelion of each body using $q = a(1-e)$ and $Q = a(1 + e)$. The results are

q (au)	Q (au)
8.769	32.017
11.8523	37.4297
6.7942	29.1538
18.9748	31.4232

Comparing these values to those in Table 5.3 yields the major HEBA class. Using Eq. (4.38) and the planetary orbital data in Table 4.1, the Tisserand value with respect to the controlling planet at perihelion can be found. The final results are

Designation	i (°)	e	a (au)	HEBA major	HEBA minor
5145 Pholus	24.6	0.570	20.393	SN	III
7066 Nessus	15.7	0.519	24.641	SE	IV
8405 Asbolus	17.6	0.622	17.974	JN	IV
10370 Hylonome	4.1	0.247	25.199	UN	IV

Example Shown are the semimajor axes and eccentricities of some selected comets. Complete this table by finding the comet class of each body according to the HEBA scheme.

Designation	a (au)	e	HEBA comet class
109P Swift-Tuttle	26.092	0.9632	
38P Stephan-Oterma	11.247	0.86002	
319P Catalina-McNaught	3.57	0.666084	
36P Whipple	4.165	0.2583	
9P Tempel	3.145	0.5096	

Solution First, the perihelion and aphelion of each comet must be found using $q = a(1 - e)$ and $Q = a(1 + e)$. The results are

q (au)	Q (au)
0.96019	51.2238
1.5744	20.9196
1.1921	5.9479
3.0892	5.2408
1.5423	4.7477

Second, the perihelion and aphelion of each comet can be compared to the values in Table 5.5 to determine each comet class. The results are

Designation	a (au)	e	HEBA comet class
109P Swift-Tuttle	26.092	0.9632	I
38P Stephan-Oterma	11.247	0.86002	SP
319P Catalina-McNaught	3.57	0.666084	SP
36P Whipple	4.165	0.2583	SP
9P Tempel	3.145	0.5096	SP

Example Shown are the semimajor axes and eccentricities of some selected comets. Complete this table by finding the comet class of each body according to the Levison scheme. The semimajor axis and inclination of the orbit of Jupiter are 5.2 au and 1.3° respectively.

Designation	a (au)	e	i (°)	Levison comet class
109P Swift-Tuttle	26.092	0.9632	113.45	
38P Stephan-Oterma	11.247	0.86002	17.981	
319P Catalina-McNaught	3.57	0.666084	15.074	
36P Whipple	4.165	0.2583	9.9345	
9P Tempel	3.145	0.5096	10.474	

Solution First, use Eq. (4.38) to find the Tisserand value with respect to Jupiter for each comet. Then, using Fig. 5.3 and the semimajor axis, determine the comet class. The results are

Designation	a (au)	e	i (°)	Levison comet class
109P Swift-Tuttle	26.092	0.9632	113.45	Halley-type
38P Stephan-Oterma	11.247	0.86002	17.981	Halley-type
319P Catalina-McNaught	3.57	0.666084	15.074	Jupiter-family
36P Whipple	4.165	0.2583	9.9345	Jupiter-family
9P Tempel	3.145	0.5096	10.474	Jupiter-family

5.5   Other Small Body Populations

Other small body populations also exist. Small bodies may be classified by semimajor axis, resonance and other properties. Table 5.6 lists other subpopulations of small bodies based on the resonance within which they are locked.

Another population of small bodies that overlaps the population of small solar system bodies is the minor planets. A **minor planet** is a body which is either a dwarf planet or a small solar system body which is not a comet. Thus, minor planets include dwarf planets, asteroids, Centaurs, and TNOs. SSSBs include comets and

Table 5.6 Other small body populations based on the resonance within which they are locked except for Cubewanos

Population	Resonance	Location (au)
Cubewanos	None	40–50
Cybeles	7:4 MMR Jupiter	3.58
Griquas	2:1 MMR Jupiter	3.27
Hildas	3:2 MMR Jupiter	3.97
Hungarias	9:2 MMR Jupiter	1.91
Plutinos	2:3 MMR Neptune	39.4
Thules	4:3 MMR Jupiter	4.30
Twotinos	1:2 MMR Neptune	47.8

Cubewanos—also known as classical Kuiper Belt Objects—are non-resonant and are not planet crossing. Except for Cubewanos, Location refers to the location of the resonance only

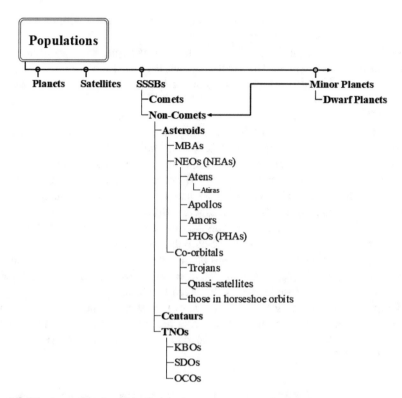

Fig. 5.5 The taxonomy of small bodies

all minor planets which are not dwarf planets. A map of the different populations of SSSBs and minor planets is shown in Fig. 5.5. Note that satellites are considered to be in a separate population.

5.6 The Effects of Planets on Small Body Populations

In a general sense, planets perturb the orbits of small bodies, causing osculating orbital parameters to vary in time. Effects on the orbits of a group of small bodies include **sculpting**, **shepherding** and **clustering**.

Sculpting creates orbital gaps in groups of small body populations. The gaps can be seen as ranges in semimajor axis in which there are relatively few bodies compared to nearby orbits.

The cause of sculpting is a planet or planet(s) in mean motion resonances with the small body population. Examples of sculpting by planets include the Kirkwood gaps in the Main Asteroid Belt (e.g. Kirkwood 1867; Wisdom 1983; Moons et al. 1998; Roig et al. 2002), gaps in the Kuiper Belt (Levison and Duncan 1993) and other TNO regions (Batygin and Morbidelli 2017).

Figure 5.1 shows gaps in the Main Asteroid Belt associated with mean motion resonances of Jupiter. Sculpting can also be applied to satellites in mean motion resonances with ring particles about a planet. For example, the planet Saturn has gaps caused by its satellites. The satellite Mimas is in a 2:1 mean motion resonance with ring particles in the Huygens Gap at the inner edge of the Cassini division (French et al. 2016).

And Saturn's moon Pan is in a 1:1 mean motion resonance with ring particles in the Encke gap (Spahn et al. 1993). Sculpting has even been used to detect unseen exoplanets in protoplanetary discs (Tabeshian and Wiegert 2016; Jang-Condell 2017). Figure 5.6 shows the gaps in Saturn's rings caused by sculpting.

Shepherding constrains a group of small bodies to orbits with semimajor axes within a certain relatively narrow range and is the opposite of sculpting. One example is the Hilda asteroids in a 3:2 MMR with Jupiter (e.g. Schubart 2007). The bodies are grouped around 3.97 au from the Sun as shown in Fig. 5.1. Shepherding also applies to satellites confining ring particles to a narrow semi-axis range. For example, Saturn's satellites Prometheus and Pandora shepherd particles in Saturn's F ring through a complex gravitational interaction (Murray and Dermott 1999; French et al. 2003) and are known as **shepherd satellites**.

Clustering is a grouping of the longitude of the ascending nodes and arguments of perihelion of small body orbits. Clustering has been used to infer the existence of an as yet undiscovered planet lying beyond the orbit of Neptune (Batygin and Morbidelli 2017).

Fig. 5.6 The planet Saturn with its rings. Note the gaps in Saturn's rings caused by sculpting. The smaller Encke gap lies near the outer edge of the rings. Inward lies the larger Cassini division. Photo courtesy of NASA

Exercises

1. Shown in the picture are the Sun along with the orbits of Mars, Jupiter and Neptune. Four regions are each labeled either A, B, C or D. Match each region with the following where they can be found.

Population	Location
Centaur	
Jupiter Trojan	
Main Asteroid Belt	
Trans-Neptunian	
Edgeworth-Kuiper Belt	
Origin of comets	

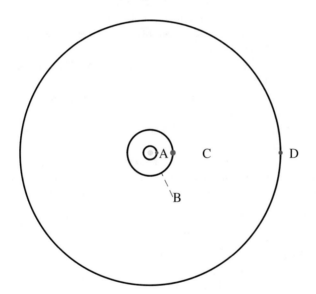

2. For each body shown, check all categories into which each object can be classified. A body may be in more than one category. Atens orbit the Sun closer to the Sun than Earth and can be irregularly shaped. Titan is a satellite of the planet Saturn. Halley's Comet crosses the orbit of more than one planet.

Body	SSSB	Minor planet	Dwarf planet
Halley's Comet			
Pluto			
An Aten			
Titan			

3. For each comet shown below, state if the comet has a period greater than 200 years, has a period less than 200 years or no longer exists.

Body	
95P Chiron	
3D Biela	
1P Halley	
85D Boethin	
C/1995 O1, Hale-Bopp	

4. Describe the typical size and composition of a Main Belt Asteroid.
5. Explain the difference between an Amor and an Apollo asteroid.
6. Explain the basic difference between a comet and an asteroid.
7. Which part of a comet is most similar to an atmosphere?
8. What does it mean for a comet to be "active"?
9. Define sublimation and explain its importance to comets.
10. Explain the main source of the activity for comets at heliocentric distances a. within 3 au and b. beyond 3 au
11. Give one reason why water-ice sublimation is probably not responsible for the activity seen on Chiron.
12. Explain why the discovery of Centaurs caused problems for the traditional definition of a comet and an asteroid.
13. What is the typical time frame that a small body spends in the Centaur region? Express in years as a power of ten.
14. Shown are the inclinations, semimajor axes and eccentricities of some selected small bodies. Complete this table by finding the major and minor HEBA classes of each body. If the object has no minor class, then leave it blank.

Designation	i (°)	e	a (au)	HEBA major	HEBA minor
2018 VG18	31.7	0.772	95.234		
2018 RR2	40.2	0.641	21.408		
2018 EZ1	29.7	0.393	16.26		
2017 YG5	23.7	0.88	62.233		
2017 WW14	6	0.747	22.747		
2017 UX51	90.5	0.747	30.111		
2016 FX59	29.6	0.06	42.462		

15. Shown are the semimajor axes and eccentricities of some selected comets. Complete this table by finding the comet class of each body according to the HEBA scheme.

Designation	a (au)	e	HEBA comet class
Halley	17.94	0.967	
21P Giacobini-Zinner	3.52	0.706	
75P Kohoutek	3.4	0.537	
Hyakutake	1165	0.9998	
81P Wild 2	3.44	0.540	

16. Shown are the semimajor axes and eccentricities of some selected comets. Complete this table by finding the comet class of each body according to the Levison scheme. The semimajor axis and inclination of the orbit of Jupiter are 5.2 au and 1.3° respectively.

Designation	a (au)	e	i (°)	Levison comet class
Halley	17.94	0.967	162.2	
21P Giacobini-Zinner	3.52	0.706	31.8	
Chariklo	15.8	0.172	23.4	
Hyakutake	1165	0.9998	124.9	
81P Wild 2	3.44	0.540	3.2	

Chapter 6
Ringed Small Bodies

The field of ringed small bodies was born in 2013 by the serendipitous discovery of two narrow rings around the Centaur Chariklo, which orbits between Saturn and Uranus (Braga-Ribas et al. 2014). Since then, rings have been detected around the TNO Haumea (Ortiz et al. 2017), and may exist around the Centaur Chiron (Ortiz et al. 2015). The existence of these ringed small bodies naturally raises several questions such as: how did the rings form? Are ringed small bodies commonplace? And what is the longevity of such rings?

Explanations for the origin of rings around small bodies include an impactor (Melita et al. 2017), a collision between an orbiting satellite and another body (Melita et al. 2017), the tidal disruption of an orbiting satellite (El Moutamid et al. 2014), ejected debris from cometary activity (Pan and Wu 2016) and the tidal disruption of the small body due to a close encounter between the small body and a giant planet (Hyodo et al. 2016).

The question of longevity becomes especially poignant in the Centaur region, where close encounters between Centaurs and giant planets are common (Horner et al. 2004) and are quite capable of severely damaging or destroying such rings.

But even if rings could survive a close encounter with a giant planet, viscous forces should widen such rings on time scales such as hundreds of years (Michikoshi and Kokubo 2017) or 100,000 years (Pan and Wu 2016), which is far shorter than the typical $\sim$10 Myr lifetime of a Centaur (Tiscareno and Malhotra 2003). However, the lifetime of rings could be extended by orders of magnitude due to stabilizing shepherd satellites (El Moutamid et al. 2014; Ortiz et al. 2015), apse alignment due to self gravity (Rimlinger et al. 2017), or resonant perturbations due a perturber or nonuniform mass distribution within Chariklo itself (Lewis and Sickafoose 2017).

If the body is active, ejected material due to activity could replenish ring particles and extend ring lifetime. This idea is reasonable given the fact that some Centaurs are known to be active (Bus et al. 1991; Jewitt 2009) and satellites of giant planets are known to be sources for planetary ring material (e.g. Burns et al. 1999; Hedman et al. 2007).

© The Author(s), under exclusive license to Springer Nature Switzerland AG 2019 93
J. Wood, *The Dynamics of Small Solar System Bodies*, SpringerBriefs
in Astronomy, https://doi.org/10.1007/978-3-030-28109-0_6

Furthermore, simulations indicate that close encounters between Chariklo or Chiron and a giant planet capable of severely damaging or destroying such rings are very rare (Araujo et al. 2016, 2018; Wood et al. 2017, 2018a; Sfair et al. 2018) and thus the rings could survive a trip through the Centaur region.

This introduces the possibility that rings around Centaurs could predate their entrance into the Centaur region. This idea is bolstered by the discovery of a ring around the dwarf planet Haumea, which lies in the Trans-Neptunian region (Ortiz et al. 2017).

However, TNOs with rings in orbits for which the dynamics are not controlled by a planet are not necessarily immune to the effects of close encounters with giant planets, as gravitational perturbations can decrease the perihelia of their orbits until they are planet crossing.

The survivability of a ring around a SSSB during a close encounter with a giant planet is related to its orbital radius (Wood et al. 2018b). Rings with smaller orbital radii are harder to perturb than rings with larger orbital radii. Sicardy et al. (2019) have shown that rings of particles within the 1:1 spin-orbit resonance of a non-axisymmetric body such as Chariklo or Haumea are absorbed by the body itself within just decades, and that rings beyond the 1:1 spin-orbit resonance are pushed beyond the location of the 1:2 spin-orbit resonance (two rotations of the body for every one orbital period).

This indicates that for such bodies with rings, the 1:2 spin-orbit resonance must exist within the Roche limit. Since the faster the rotation of the body the closer the 1:2 spin-orbit resonance is to the body, bodies with faster rotation periods are favored to have rings over bodies with slower rotation periods.

To learn more about the origin of such rings and the feasibility of ring survival in the Centaur region, it is beneficial to study each of these three bodies in more detail and to investigate close encounters between ringed small bodies and giant planets.

6.1   Chariklo

Chariklo is a Centaur orbiting between Saturn and Uranus with a semimajor axis of 15.8 au, an eccentricity of 0.172 and an inclination of 23.4°. Various values for the mean radius of Chariklo have been reported and range from 119 km (Fornasier et al. 2014) to 137 km (Altenhoff et al. 2001).

Chariklo is thought to have a non-spherical shape with an major to minor ellipsoid axis ratio of 1.1 (Fornasier et al. 2014). Its surface is believed to be composed of mostly refractory material and water ice (Groussin et al. 2004).

Chariklo's density is poorly known. Braga-Ribas et al. (2014) suggest a bulk density of $1000 \, \mathrm{kg \, m^{-3}}$. Using this density and reported size range, this puts the mass of Chariklo somewhere in the range 10^{18}–10^{19} kg.

The ring structure of Chariklo consists of two separate rings. An inner ring of radius 391 km and width 7 km and an outer ring of radius 405 km and width 3 km. The rings are believed to be composed of mostly silicates (compounds of silicon

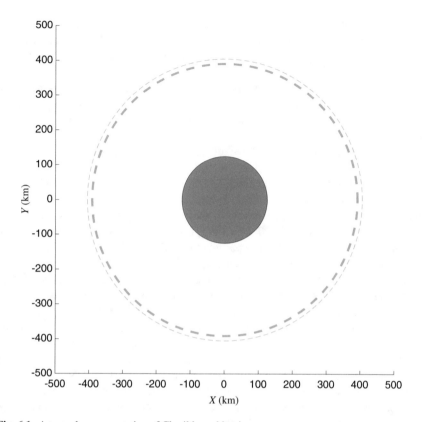

Fig. 6.1 A to-scale representation of Chariklo and its rings

and oxygen) with water ice, tholins (organic compounds formed by solar ultraviolet irradiation or cosmic rays) and amorphous carbon (Duffard et al. 2002).

A to-scale representation of Chariklo and its rings is shown in Fig. 6.1. At this time, Chariklo is not believed to be active, and no activity has been detected since its discovery. However, it is not known if Chariklo has been active in the past.

6.2 Chiron

Chiron's physical properties have been difficult to ascertain due to interference from circum-nuclear material, Chiron's non-spherical shape and cometary activity. Nevertheless, a concerted effort has been taken to obtain properties such as size, shape, and density (Fornasier et al. 2013; Ortiz et al. 2015). Proposed radii for Chiron range from 71 to 186 km (Sykes and Walker 1991; Groussin et al. 2004; Ortiz et al. 2015).

Chiron's bulk density cannot be established due to the large uncertainty in its size. A density range of 500–1000 kg m^{-3} has been proposed for Chiron (Meech et al. 1997). Assuming a spherical body, this places the mass of Chiron somewhere on the order of 10^{17}–10^{19} kg.

Chiron is believed to be porous and comprised of a mixture of dust and ices such as CN, CO, CO_2, and water ice (Stern 1989; Luu and Jewitt 1990; Meech and Belton 1990; Bus et al. 1991; Prialnik et al. 1995; Womack and Stern 1995; Capria et al. 2000).

Since its discovery, observers have reported sporadic outbursts or changes in the brightness of Chiron perhaps related to cometary activity (Bus et al. 1988, 2001; Hartmann et al. 1988, 1990; Luu 1993; Lazzaro et al. 1996, 1997; Silva and Cellone 2001; Duffard et al. 2002). In addition to sporadic changes in brightness, Chiron also displays periodic activity which reaches a peak about every decade. This may or may not be real (Luu and Jewitt 1990; Luu 1993; Lazzaro et al. 1996) but is unrelated to the rotation of Chiron, which has a period of only 5.92 h (Bus et al. 1989).

Because of this unusual activity, an organized global observation campaign was started to follow Chiron through its perihelion passage in the mid-1990s (Stern 1995), and this led to Chiron becoming the first object ever to receive both an asteroidal and a cometary designation: 2060 Chiron and 95P/Chiron.[1]

Since Chiron displays activity at heliocentric distances beyond 3 au (Stern 1989; Meech and Belton 1990; Hartmann et al. 1990; Luu and Jewitt 1990; Meech et al. 1997), water-ice sublimation is not responsible for the activity. One likely source of this activity is the outgassing of highly volatile ices such as CN, CO or CO_2, which escape due to thermal sublimation or the transformation of amorphous water ice to its crystalline form (Stern 1989; Meech and Belton 1990; Prialnik et al. 1995; Lazzaro et al. 1997; Capria et al. 2000).

After the discovery of rings around Chariklo, a reanalysis of star occultation data showed that the circum-nuclear material around Chiron could be interpreted as rings with a mean radius of 324 ± 10 km (Ortiz et al. 2015). An **occultation** occurs when one celestial object passes in front of another.

6.3 Haumea

Haumea is a dwarf planet and is the most massive of the ringed small bodies having a mass of 4.006×10^{21} kg, which may be three orders of magnitude larger than the mass of Chiron or Chariklo. Its mass is better constrained due to Haumea's two moons named Hi'iaka and Namaka.[2] Its surface is believed to be comprised of water ice with a dark spot (Lacerda et al. 2008; Thirouin et al. 2016) and no atmosphere (Ortiz et al. 2017).

The shape of Haumea is far from spherical, having a shape based on three axes known as a triaxial ellipsoid. The size of Haumea is based on thermal models. One

---

[1] http://www.minorplanetcenter.net/iau/lists/PeriodicCodes.html (accessed 23 Dec. 2016).

[2] https://planetarynames.wr.usgs.gov/Page/Planets#DwarfPlanets.

result for the length of its three axes is $2000 \times 1500 \times 1000$ km (Rabinowitz et al. 2006). Ortiz et al. (2017) report that Haumea's major and minor axes have lengths of 1704 ± 4 km and 1138 ± 26 km, respectively.

This shape is due to its very fast rotational period of 3.9 h (Rabinowitz et al. 2006; Lacerda et al. 2008). Haumea's density is in dispute. Rabinowitz et al. (2006), Lellouch et al. (2010), Lockwood et al. (2014) and Thirouin et al. (2016) report a density of at least 2500 kg m^{-3}, which is larger than a typical density of a TNO (Thirouin et al. 2016), though other TNOs such as 2002 GH$_{32}$ and 2003 UZ$_{413}$ have similar high densities (Perna et al. 2009; Thirouin et al. 2016). By contrast, Ortiz et al. (2017) report an upper limit on Haumea's density of 1885 kg m^{-3}, which if correct agrees more with other typical TNO densities.

The ring is believed to orbit in the equatorial plane of Haumea with an orbital radius of 2287 km and a width of 70 km. The ring exhibits 3:1 spin-orbit coupling with Haumea's rotation (Ortiz et al. 2017). The ring may be the result of a collision as Haumea is a member of the only known TNO collisional family though this is far from certain. Besides Haumea and its two moons, the family has ten other members (Rabinowitz et al. 2006; Thirouin et al. 2016).

6.4   The Detection of Rings

The rings of Saturn have been known since 1610 and can be seen from Earth in even small telescopes. However, rings around other bodies in our solar system are too faint and/or too close to the body to be seen directly from Earth-based telescopes. These rings must be detected using a method that does not involve direct telescopic observation.

One method is to monitor the light intensity from a star during an occultation by a ringed body. Then, as the rings around the body pass in front of the star, the rings block some of the starlight, causing a decrease in the intensity of the visible light detected from the star.

A graph of intensity vs. time is known as a **light curve**. Given the circular or elliptical shape of rings, any particular ring may cause up to two dips in the light curve. One dip occurs when the ring passes in front of the star, ahead of the small body (the **ingress**). The second dip occurs after the small body has occulted the star (if it does), when the part of the ring trailing behind the body blocks some of the starlight (the **egress**). Viewing an occultation like this from different locations on Earth allows the shape of the small body to be inferred and the pole orientation to be estimated.

This technique was used to discover the rings of Uranus (Planetary Laboratory et al. 1978), Chariklo (Braga-Ribas et al. 2014), Haumea (Ortiz et al. 2017) and possible rings around Chiron (Elliot 1995; Bus 1996; Ortiz et al. 2015). Figure 6.2 shows the light curve during the occultation of a star by a hypothetical small body. The large dip in the center is caused by the small body itself. The two smaller dips are caused by a ring.

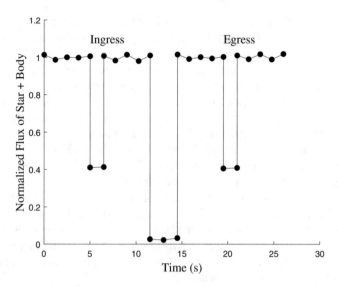

Fig. 6.2 The light curve during the occultation of a star by a small body. The large dip in the center is caused by small body itself. The two smaller dips are caused by a ring

Valuable information can be gained by analyzing light curves like the one shown in Fig. 6.2, but other factors must be considered before pertinent information can be gained. If the relative speed of the body is known, it can be used in conjunction with the time interval for a dip to determine a ring width.

A range of possible ring orbital radii can also be found. To do this, star occultation data can be analyzed to determine the locations in the sky relative to the body at which the dips occurred. It is beneficial if the occultation is viewed from multiple locations, as this can provide multiple data points. Simplifying assumptions can be made, such as assuming the rings are circular and lie in the equatorial plane of the body.

Another important factor to consider when using this technique is the angle at which the rings are being viewed. The **aspect angle** is defined as the angle between the rotation axis of the body and the line of sight direction of the observer. If the aspect angle is known or even if just its range is known, this can be used to find possible coordinates of the poles of the body.

6.5 Measuring the Severity of Close Encounters Between Ringed Small Bodies and Planets

When a ringed small body that's not a satellite has a close encounter with a planet, it is in a parabolic or hyperbolic orbit about the planet. As the ringed body flies by the planet, the orbits of the ring particles about the small body are perturbed due to tidal forces and this alters the osculating orbital parameters of ring particle orbits.

Fig. 6.3 An example of a close encounter between a ringed small body and a planet. Before the close encounter at point A, the ring particles are in circular orbits about the small body and are in the plane of the body's orbit about the planet. At point B, the small body is at the close encounter distance, d_{min}, from the planet. The ring particle orbits are now ellipses. At point C, the close encounter has ended

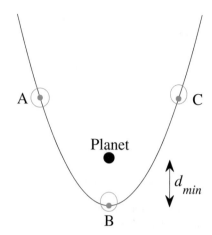

One way to measure the severity of a close encounter is to measure the change in eccentricity, Δe_{max}, of the most perturbed ring particle orbit. The greater the Δe_{max}, the greater the severity of the encounter.

Since the eccentricity of a ring particle orbit will only be perturbed by gravitational forces in the plane of the orbit (Murray and Dermott 1999), it is therefore logical to consider close encounters in which the rings lie in the plane of the ringed body's parabolic or hyperbolic orbit about the planet, since rings inclined to the orbital plane would be relatively harder to perturb.

Figure 6.3 shows a close encounter between a hypothetical ringed body and a planet, with the rings being in the plane of the small body's orbit. At point A, the ring particles are in circular obits. At point B, the ringed body is at its distance of closest approach to the planet, d_{min}, and the ring particle orbits have already begun to be perturbed. At point C, the close encounter is over, and the ring particles now clearly orbit in ellipses due to the increase in eccentricity of the ring particle orbits.

To measure the severity of this encounter, the change in eccentricity of each ring particle orbit is found, and then Δe_{max} is used to measure the severity.

Given all other factors the same, Δe_{max} increases with decreasing d_{min} due to the increasing tidal forces at smaller distances as shown in Eq. (4.7). Thus, given all other factors the same, a close encounter with a smaller d_{min} is more severe than one with a larger d_{min}.

If the effects of v_{∞} are considered to be negligible, then a severity scale based on d_{min} alone can be created. Certain critical distances naturally lend themselves as boundaries for encounters of different severity. These are: the Hill Radius, the **tidal disruption distance**, R_{td}, and the **Roche limit**, R_{roche}.

The tidal disruption distance is that close encounter distance at which the outermost ring particle is at its Hill radius relative to the small body. That is, if the close encounter distance is just within the tidal disruption distance, then the outermost ring particle is longer within the Hill sphere of the small body relative to the planet and no longer orbits it.

The equation for the tidal disruption distance can be found by replacing the mass of the Sun with the mass of the planet, setting the Hill radius equal to the orbital radius of the outermost ring particle, r, and solving (4.35) for the radial distance (which is the tidal disruption distance). The result is

$$R_{td} \approx r \left(\frac{3M_p}{m_s} \right)^{\frac{1}{3}}$$

(6.1)

(Agnor and Hamilton 2006; Philpott et al. 2010). Here, m_s is the mass of the small body. Thus, the tidal disruption distance represents the distance within which the rings can be removed from the small body.

Example Find the tidal disruption distance between Neptune of mass 1.03×10^{26} kg and a Chariklo of mass 8×10^{18} kg and ring orbital radius 405 km.

Solution Using Eq. (6.1), the tidal disruption distance is

$$R_{td} \approx r \left(\frac{3M_p}{m_s} \right)^{\frac{1}{3}}$$

$$R_{td} \approx 405 \left(\frac{3 \times 1.03 \times 10^{26}}{8 \times 10^{18}} \right)^{\frac{1}{3}}$$

$$R_{td} \approx 1.37 \times 10^5 \text{ km}$$

The Roche limit is the separation distance between two bodies—in this case a small body and a planet—at which the tidal force on an infinitesimal point mass $m_{particle}$ on the surface of the first body at the point closest to the second body just equals the gravitational force of attraction of the first body on $m_{particle}$ (see point A in Fig. 4.5).

Given a spherical small body of radius R_s, whose center of mass is at a distance R_{roche} from the center of mass of a planet. The tidal force at the point in question is given by Eq. (4.7), and the gravitational force of attraction of the small body on $m_{particle}$ is given by Eq. (4.5). These two forces are in opposite directions and are equal.

Equating the two forces yields:

$$\frac{2G_c M_p m_{particle} R_s}{R_{roche}^3} = \frac{G_c m_s m_{particle}}{R_s^2}$$

(6.2)

Equation (6.2) can be solved for the Roche limit, R_{roche}. The result is:

$$R_{roche} = R_s \left(\frac{2M_p}{m_s} \right)^{\frac{1}{3}}$$

(6.3)

Equation (6.3) can be rewritten in terms of the densities of the planet and small body ρ_p and ρ_s respectively. The result is:

$$R_{roche} = R_p \left(2 \frac{\rho_p}{\rho_s} \right)^{\frac{1}{3}} \tag{6.4}$$

R_p is the physical radius of the planet. Ignoring other forces, just within the Roche limit, the small body can be torn apart by tidal forces.

Example Find the Roche limit between Neptune of mass 1.03×10^{26} kg and a Chariklo of mass 8×10^{18} kg and mean radius 125 km.

Solution Using Eq. (6.3), the Roche limit can be found. The result is

$$R_{roche} = R_s \left(\frac{2M_p}{m_s} \right)^{\frac{1}{3}}$$

$$R_{roche} = 125 \left(\frac{2 \times 1.03 \times 10^{26}}{8 \times 10^{18}} \right)^{\frac{1}{3}}$$

$$R_{roche} = 3.69 \times 10^4 \text{ km}$$

In addition to the Hill radius, tidal disruption distance and Roche limit, Wood et al. (2017) introduced a new critical distance applicable to ringed small bodies only. This new critical distance, named the **ring limit**, R, is defined as the close encounter distance at which the effect of the close encounter on a ring is just "noticeable".

The effect is just noticeable if $\Delta e_{max} = 0.01$ (Araujo et al. 2016). Table 6.1 shows a scale used to gauge the severity of a close encounter between a ringed small body and a planet.

The severity scale shows that if a close encounter is Severe or Extreme, then a ringed small body could potentially lose its ring in one close encounter.

Table 6.1 A scale for gauging the severity of close encounters between a ringed small body and a planet based on the close encounter distance, d_{min}

d_{min} Range	Severity
$d_{min} \geq R_H$	Very low
$R \leq d_{min} < R_H$	Low
$R_{td} \leq d_{min} < R$	Moderate
$R_{roche} \leq d_{min} < R_{td}$	Severe
$d_{min} < R_{roche}$	Extreme

R_H, R, R_{td} and R_{roche} are the Hill radius of the planet with respect to the Sun, ring limit, tidal disruption distance and Roche limit respectively

Table 6.2 Examples of the critical distances for close encounters between each giant planet and a Chariklo of mass 8×10^{18} kg, density $1000 \, \text{kg m}^{-3}$ and ring orbital radius 405 km for low v_∞

Quantity	Jupiter	Saturn	Uranus	Neptune
R_{roche} (R_{td})	0.277	0.273	0.269	0.269
R_{td} (R_{td})	1	1	1	1
R (R_{td})	1.8	1.8	1.8	1.8
R_H (R_{td})	146	269	540	847

Table 6.2 shows examples of the critical distances for close encounters between each giant planet and a Chariklo of mass 8×10^{18} kg, density $1000 \, \text{kg m}^{-3}$ and ring orbital radius 405 km for low v_∞.

6.6   The Ring Limit

Wood et al. (2018b) found that the ring limit was a function of the v_∞ of the small body's orbit about the planet, but that the lower bound of R was approximately $1.8 R_{td}$ (Sfair 2018) for any planet-small body mass pair as long as $M_p \gg m_s$.

This lower bound value occurs when $v_\infty = 0$, that is for a parabolic orbit. The top diagram in Fig. 6.4 shows the ring limit as a function of v_∞ for close encounters between Saturn and a small body with the mass of Pluto $= 1.309 \times 10^{22}$ kg for ring orbital radii in the range 25,000–100,000 km found using a numerical integration.

As the diagram shows, when the ring limit is expressed in units of tidal disruption distances, it decreases with decreasing v_∞ and reaches a lower bound when $v_\infty = 0$.

Going from the lowest to highest contour, the ring orbital radius increases from 25,000 km to 100,000 km usually in increments of 5000 km. It can be seen that as v_∞ approaches zero, the contours nearly converge into one and all have a lower bound of approximately $1.8 R_{td}$.

This lower bound is only slightly dependent on the planet mass (Wood et al. 2018b). As an example of this, the bottom diagram of the figure shows ring limits for the same small body with a ring orbital radius of 50,000 km for close encounters with Neptune. The lower bound rounds off to $1.8 R_{td}$.

Though the contour shape for Neptune is different than that of Saturn's. For low and intermediate v_∞, the contour has a similar shape to those of Saturn, but note how the plot plateaus at high v_∞. This is something that the contours of Saturn do not do in this v_∞ range.

The top diagram in Fig. 6.5 shows a histogram of ring limits for close encounters between the planet Saturn and small bodies each with the mass of Pluto and a unique ring orbital radius from the range 25,000 km to 100,000 km, usually in increments of 5000 km for the case where $v_\infty = 0$ (parabolic orbits). The histogram shows the values clustered around $1.8 \pm 0.00396 R_{td}$, which is the mean value.

The bottom diagram in Fig. 6.5 shows a plot of the same ring limit lower bounds versus ring orbital radius. It can be seen that for any ring orbital radius, the ring limit lower bound is approximately $1.8 R_{td}$.

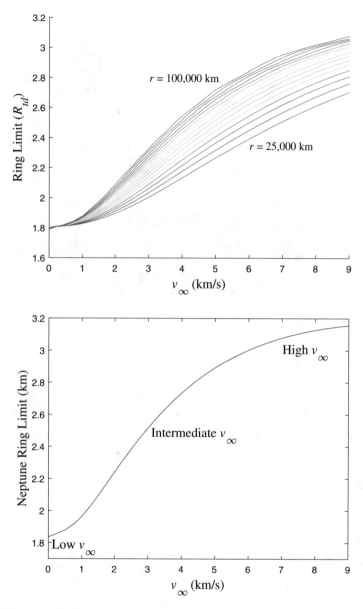

Fig. 6.4 Top—ring limits for close encounters between a small body with the mass of Pluto and the planet Saturn. Bottom—ring limits for the same small body with a ring orbital radius of 50,000 km for close encounters with Neptune

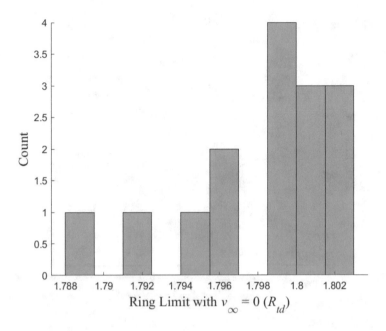

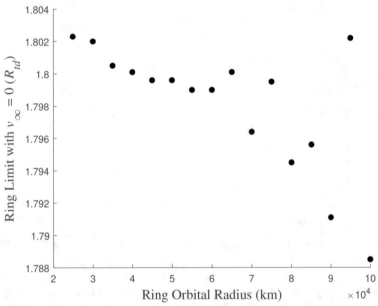

Fig. 6.5 Ring limits for close encounters between the planet Saturn and small bodies each with the mass of Pluto and a unique ring orbital radius from the range 25,000 km to 100,000 km, usually in increments of 5000 km for the case where $v_\infty = 0$ (parabolic orbits). Top—as a histogram. Bottom—vs ring orbital radius

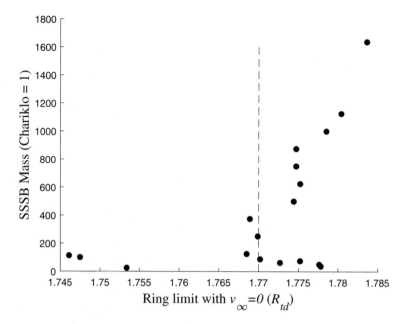

Fig. 6.6 Ring limit lower bounds ($v_\infty = 0$) for close encounters between Jupiter and SSSBs with masses in the range 25–1636 Chariklo masses. The dashed line is the mean value of $1.77 \pm 0.01 R_{td}$

Figure 6.6 shows ring limit lower bounds ($v_\infty = 0$) for close encounters between Jupiter and a variety of SSSBs with masses in the range 25–1636 Chariklo masses where one Chariklo mass is 8×10^{18} kg. The mean ring limit lower bound of $1.77 \pm 0.01 R_{td}$ rounds up to $1.8 R_{td}$.

This means that if $d_{min} <\approx 1.8 R_{td}$, the effect of the close encounter on the ring is always noticeable regardless of the ring orbital radius, SSSB mass or giant planet mass. Thus, the value of $1.8 R_{td}$ lends itself as an important critical distance. For low v_∞, velocity effects can be ignored and R can be set to a constant value of $1.8 R_{td}$ for simplicity.

For intermediate v_∞, a numerical integration could be used to create contours such as those shown in Fig. 6.4 for the particular small body-planet close encounter in question over a desired v_∞ range. Then, interpolation could be used to find the value of a ring limit for a single value of v_∞ within the range.

If the ring limit in units of R_{td} is desired over a range of v_∞ values then the data could be fit to a function using linear regression or a power regression of the form

$$R = Cv_\infty^\delta \tag{6.5}$$

Though, this method may not be as accurate as interpolation. As an example of using this method, Table 6.3 contains sample ring limit data for close encounters between Saturn and a small body with the mass of Pluto and ring orbital radius of

Table 6.3 Sample ring limit data for close encounters between Saturn and a small body with the mass of Pluto and ring orbital radius of 25,000 km

v_∞ (km/s)	$R(R_{td})$	v_∞ (km/s)	$R(R_{td})$
9	2.7042	4.5	2.2003
8.9	2.6954	4.25	2.167
8.75	2.6825	4	2.1335
8.5	2.6603	3.9	2.1202
8.25	2.6373	3.75	2.1003
8	2.6134	3.5	2.0675
7.9	2.6036	3.25	2.0353
7.75	2.5888	3	2.0041
7.5	2.5632	2.9	1.9919
7.25	2.5371	2.75	1.9743
7	2.5101	2.5	1.9461
6.9	2.4991	2.25	1.9198
6.75	2.4821	2	1.896
6.5	2.4536	1.9	1.8872
6.25	2.424	1.75	1.8746
6	2.3938	1.5	1.8562
5.9	2.3816	1.25	1.8405
5.75	2.363	1	1.8276
5.5	2.3316	0.9	1.8234
5.25	2.2995	0.75	1.8178
5	2.2668	0.5	1.8106
4.9	2.2535	0.25	1.8064
4.75	2.2337		

25,000 km. When fit to Eq. (6.5), $C = 1.83$ and $\delta = 0.147$. Thus, in this particular case, the ring limit in units of R_{td} could be approximated over this range of v_∞ values by $R = 1.83v_\infty^{0.147}$.

Finally, for high values of v_∞, the ring limit could be found using interpolation or be set to the largest value of R found over the range in question. The latter method works best if the contour plateaus as shown in the bottom diagram of Fig. 6.4. If the contour does not plateau and the v_∞ lies outside the range of study, then extrapolation could be used.

In summary, the ring limit can be found using these techniques

- low v_∞—set the ring limit to a constant value of $1.8R_{td}$.
- intermediate v_∞—use interpolation.
- high v_∞—use interpolation or set the ring limit to the largest value of R found.
- v_∞ outside the range studied—set the ring limit to the largest value of R found or use extrapolation.

The experimenter determines if a v_∞ value is low, intermediate or high.

6.7   The Future Study of Ringed Small Bodies

The field of ringed small bodies is a relatively new subfield of astronomy. Rings are known to exist around Chariklo and Haumea; and there is a possibility that rings exist around Chiron. There are major questions that are in need of answering for this very young field.

For example, are rings around small bodies commonplace or rare? How stable are rings around small bodies? Do shepherd moons exist around Chariklo and Haumea? Are all small body rings created in the same way or are multiple ring creation mechanisms at work?

The pioneering work of Wood et al. (2018b) has laid a foundation for the study of any rings that are found around small bodies in the future by demonstrating practical techniques, providing a severity scale for close encounters and providing data that can be extrapolated or interpolated to determine ring limits for close encounters between giant planets and other ringed small bodies.

This work could be built upon by simulating close encounters between small bodies and giant planets for small body masses outside the mass range used in this work. Hopefully, future research will corroborate their results. Their work may also have applications in other fields such as the study of circumstellar disks and planetary systems that have encounters with other stars.

Information learned from the study of ringed small bodies can be applied to theoretical studies of hypothesized small body populations in other solar systems. This scenario becomes more likely as the number of discovered exoplanets continues to increase.

Our knowledge in this field could grow significantly with an increase in the number of probe missions to small bodies. A probe mission to Chariklo such as the one proposed by Bouchard et al. (2018) could determine if it has shepherd moons to help stabilize its rings as has been proposed (El Moutamid et al. 2014; Ortiz et al. 2015). It could also help determine if the gap between Chariklo's rings is caused by a moon in the gap, or outside the gap. If it is the former, then the gap would be formed in the same way that the moon Pan helps create Saturn's Encke gap (Porco et al. 2005).

If the latter, then a moon outside the gap would create it using sculpting. Of course, the gap could be caused by neither of these mechanisms. A probe mission to Chiron could finally determine conclusively if the detected circum-nuclear material around this body is a ring, an arc, jets or something else entirely.

Probe missions would also lead to better measurements of rotation speeds of small bodies with rings and allow for a better determination of the locations of spin-orbit resonances, which have consequences for ring evolution.

Now that we know that Haumea is the only known ringed dwarf planet with moons and is a member of a collisional family, this makes it an object of interest. Since space probes are now beginning to visit the Trans-Neptunian region, this author proposes that Haumea be selected as the next TNO to be visited by a probe.

Innovations by the SpaceX company are making probe missions more affordable. The company recently launched a car into space and demonstrated again the reusability of its booster rockets.

As costs go down, missions to the outer solar system will become more affordable, which hopefully will lead to more probe missions and a significant enhancement of our knowledge of small solar system bodies.

The Gaia space probe is in the process of measuring "the positions of about one billion stars both in our Galaxy and other members of the Local Group, with an accuracy down to 24 microarcseconds".[3]

Knowing the positions of these stars more accurately than ever before will reduce the uncertainty in the paths that shadows take across the Earth during an occultation event. This in turn may lead to more observations of stellar occultations and to the discovery of more ringed small bodies.

Thus, we may be on the verge of tremendous growth in this field. In fact, given that the number of Centaurs with diameters larger than 1 km is believed to exceed 44,000 (Horner et al. 2004), there could exist a large undiscovered population of ringed Centaurs.

Numerical integrations will continue to play an important role in the study of ringed small bodies. Numerical integrations of close encounters between each planet and Chariklo and Chiron should be carried out to study ring stability for v_∞ ranges outside that used in Wood et al. (2017, 2018a,b) of $0 \leq v_\infty \leq 9$ km/s. These integrations could involve the small body and massless ring particles both with and without shepherd moons.

A study of close encounters between each planet and Haumea for an appropriate v_∞ range should also be done. Furthermore, integrations of initially circular or non-circular rings could be done to determine a time frame within which stabilizing factors such as shepherd moons and self-gravitating rings can heal the damage done to the ring structure by the close encounter.

The average time between close encounters for each ringed body should be found to determine if rings can heal themselves via shepherd moons (El Moutamid et al. 2014) or self-gravitating rings (Rimlinger et al. 2017) before another destructive encounter occurs.

Integrations of actual collisions between small bodies could also be simulated using smooth particle hydrodynamics (Canup 2005; Benz et al. 2008) to study ring formation. These would involve collisions between the small body itself and bodies of various masses or collisions between satellites about the small body. Admittedly, these integrations would be time intensive given the large number of separate particles involved.

Integrations of possible formation scenarios of the Haumea system including its ring are also of interest. The exact ring formation mechanism is far from clear. Though Haumea is a member of the only known Tran-Neptunian collisional family,

---

[3]http://sci.esa.int/gaia/47354-fact-sheet/ (accessed April 6, 2019).

this does not necessarily mean that its rings were formed by a collision, especially since the ring also lies within Haumea's Roche limit (Ortiz et al. 2017).

It is also interesting that the orbital radius of the ring around Haumea is an order of magnitude larger than those of Chariklo or Chiron. Though Haumea currently does not cross the orbit of any planet, the possibility that Haumea has crossed the orbits of planets in the past cannot be excluded.

The larger members of the Trojan populations of Jupiter and Neptune should also be checked for rings. This need not involve a probe necessarily but instead could be conducted using stellar occultations. Also, integrations of potential ring-bearing Trojans would make an interesting study.

Then the effect of other asteroids on rings could be studied. This is especially true in the Jupiter Trojan region where significantly more asteroids have been discovered. Though, many thousands of Neptune Trojans could remain undiscovered (Sheppard and Trujillo 2006; Lykawka et al. 2011).

For Jupiter Trojans, possible candidates include 624 Hektor, 911 Agamemnon, and 1437 Diomedes, which have diameters comparable to those of Chariklo and Chiron.[4] For Neptune Trojans, possible candidates include 2001 QR322, 2011 HM102, 2006 RJ103, 2007 VL305, (316179) 2010 EN65, 2011 WG157, and 2013 KY18.

Whether by ground or space mission, our knowledge of ringed small bodies is sure to increase. As our knowledge of this young field grows, many more ringed small bodies may be found. One day, their number may rival that of the currently known Centaurs themselves. Then, perhaps questions involving the nature and origin of the rings will be answered.

Exercises

1. Given the small ringed body Chiron, which has a mass of 0.00000051 Earth masses, and Saturn, which has a mass of 95 Earth masses. If the ring orbital radius of Chiron's ring is 324 km, what is the tidal disruption distance in kilometers between Chiron and Saturn?
2. Given a small body with a density of $1000 \, \mathrm{kg \, m^{-3}}$. If the planet Saturn has a density of $687 \, \mathrm{kg \, m^{-3}}$ and a mean radius of 9.45 Earth radii, what is the Roche limit between this body and the planet Saturn?
3. Given initially circular orbits for ring particle orbits about a small body, what is required in order for a close encounter between a ringed small body and a planet to have a noticeable effect on the ring?
4. How is a ring limit defined? Given three small ringed bodies of equal mass of $8 \times 10^{18} \, \mathrm{kg}$. The ring orbital radius for each body is Body A: 324 km, Body B: 405 km and Body C: 2200 km. If each small body has a close encounter with the

---

[4]https://ssd.jpl.nasa.gov/sbdb.cgi (accessed Feb. 5, 2018).

same planet at the same v_∞, then which small body has the largest ring limit if the ring limit is measured in kilometers? If the ring limit is measured in units of tidal disruption distances, then according to Fig. 6.4 which body would have the largest ring limit?

5. When the ring limit is expressed in units of tidal disruption distances, what is the approximate minimum value the ring limit can obtain, and for what type of close encounter orbit does this apply to?

6. Given a small body with the mass of Pluto (0.00220 Earth mass), density 1860 kg m^{-3} and ring orbital radius of 50,000 km that has a close encounter with Saturn, of mass 95 Earth masses, density 687 kg m^{-3}, and orbit semimajor axis 9.55 au. Given the mass of the Sun is 332,525 Earth masses. Express the Hill radius of Saturn and Roche limit in units of tidal disruption distances.

7. Given a small body with the mass of Pluto. Given a set of close encounters between the body and Saturn with the v_∞ and distance of closest approach to Saturn, d_{min}, shown for each encounter, rank the severity of each close encounter as very low, low, moderate, severe or extreme using values obtained in the previous problem and Fig. 6.4.

v_∞(km/s)	$d_{min}(R_{td})$
0	2.5
3	1.5
5	0.003
9	30

8. Shown in the following table is ring limit data for close encounters between Neptune and a small body with the mass of Pluto and ring orbital radius of 50,000 km. Fit this data to a power law of the form $R = Cv_\infty^\delta$. State the values of C, δ and the regression coefficient. Plot the ring limit R vs v_∞ and compare the shape of this contour to those of Saturn shown in Fig. 6.4.

v_∞ (km/s)	$R(R_{td})$	v_∞(km/s)	$R(R_{td})$
9	3.156	4.5	2.8191
8.9	3.1539	4.25	2.7778
8.75	3.1505	4	2.7335
8.5	3.1441	3.9	2.7144
8.25	3.1365	3.75	2.6853
8	3.1274	3.5	2.6331
7.9	3.1234	3.25	2.5769
7.75	3.117	3	2.5168
7.5	3.1057	2.9	2.4914
7.25	3.0929	2.75	2.4525
7	3.0785	2.5	2.3849
6.9	3.0722	2.25	2.314

(continued)

v_∞ (km/s)	$R(R_{td})$	v_∞ (km/s)	$R(R_{td})$
6.75	3.0621	2	2.2406
6.5	3.0442	1.9	2.2109
6.25	3.0247	1.75	2.1665
6	3.0026	1.5	2.0936
5.9	2.9932	1.25	2.0246
5.75	2.9784	1	1.9636
5.5	2.9523	0.9	1.9422
5.25	2.923	0.75	1.914
5	2.8915	0.5	1.8782
4.9	2.8781	0.25	1.8567
4.75	2.8567		

9. Repeat the previous question, but this time use a linear regression over the range $0.25 \le v_\infty \le 4$ km/s. Compare the regression coefficient to that of the previous problem and comment on which fit is more accurate.

Answer Key

Chapter One

1. A comet will move relative to the stars as the weeks go by, but the stars won't. The comet may also have a different shape than a star and will eventually disappear.
2. as a meteoroid
3. If Pluto could clear the neighborhood of its orbit, it would be a planet.
4. Eros is not a dwarf planet because it is not spherical and has not cleared the neighborhood of its orbit.
5. Haumea has a mass of 6.71×10^{-4} of an Earth mass and a radius 0.0781 of an Earth radius.
6. This reasoning is flawed because dwarf planets orbit the Sun, are smaller than planets and yet are not SSSBs, because by definition a small body of the solar system cannot be a dwarf planet.

Chapter Two

1. No, not necessarily. To be a PHO, the MOID has to be within 0.05 au. There's not enough information given to determine if it is a PHO. The MOID of this body is not necessarily 0.0787 au. The MOID only measures the minimum distance between any two points between two orbits and could be less than 0.0787 au. The MOID value is not necessarily the distance of closest approach between two bodies.
2. Because in the inner solar system, ice on the body would sublimate. This would spread large amounts of debris throughout the inner solar system which could block much sunlight from reaching Earth or collide with Earth.

© The Author(s), under exclusive license to Springer Nature Switzerland AG 2019
J. Wood, *The Dynamics of Small Solar System Bodies*, SpringerBriefs
in Astronomy, https://doi.org/10.1007/978-3-030-28109-0

3. Comets could have collided with Earth over eons of time and delivered much water to Earth.

4. The reason could be to protect Earth, to learn more about the formation of our solar system, or to detect unseen planets. Answers will vary on the journal article or website.

5. Computer simulations have the advantage of being able to study objects in the past or future over time intervals far beyond the lifetime of a human. It is relatively inexpensive to do compared to a probe study. Probes allow for the study of an actual object instead of a computer generated version of an object. Therefore, probes are able to determine important properties such as composition, albedo, shape and density that a simulation cannot determine.

6. A test particle is massless while a speck of dust in space has some mass.

7. Answers will vary. Possible answers include MERCURY, SWIFT and Rebound.

8. The observer should not view a SSSB at low altitudes because atmospheric turbulence tends to degrade telescopic images. The body is best viewed when at maximum altitude above the horizon to reduce atmospheric effects. This body has not crossed its meridian yet and so has not reached maximum height. Furthermore, since this asteroid's orbit lies outside of Earth's, has relatively low eccentricity and does not cross Earth's orbit, it is best viewed at opposition or when the Sun is opposite the asteroid in the sky. The asteroid is too close to Sun to be at opposition and so this is not the best time to view it.

9. Tons of debris in the form of dust particles from asteroids, comets and other sources enters Earth's atmosphere each day. These particles can be collected and returned to Earth by balloons or aircraft in the upper atmosphere. Furthermore, if asteroid debris survives its trip through Earth's atmosphere, it can simply be picked up off the ground.

10. the semimajor axis and eccentricity of each clone would be

Clone	a (au)	e
1	15.859	0.172
2	15.859	0.175
3	15.859	0.178
4	15.864	0.172
5	15.864	0.175
6	15.864	0.178
7	15.869	0.172
8	15.869	0.175
9	15.869	0.178

11. The number of clones would be $7 \times 7 \times 7 = 343$

Chapter Three

1. Accurately predicts future celestial events and is changeable.
2. The Ptolemaic model was aesthetically correct in explaining retrograde motion and diurnal motion because it could explain the motion based on appearances, but it was not physically correct because the physical motions used to explain these phenomena did not actually exist. The Copernican model was both aesthetically and physically correct in describing these motions because it made use of physical motions that actually existed to explain these phenomena.
3. Deferent—larger sphere, epicycle—smaller sphere, eccentric—Earth off center.
4. The equant was a point off center of the deferent opposite Earth. An observer at the equant would see the center of a planet's epicycle move at a uniform speed.
5. Because the center of the motion is neither the Earth nor the Sun, but rather a geometric point called the equant at which there is nothing.
6. The Ptolemaic model explained retrograde motion using epicycles moving along a deferent while a planet spun along the rim of the epicycle. The center of the epicycle moved along the rim of the deferent carrying the planet along with it. Occasionally, an observer on Earth would see the planet physically move backwards from its regular motion due to its motion along the epicycle. Copernicus explained retrograde motion as an illusion due to the passing of planets in their orbits about the Sun. In this model, a planet would only appear to move backwards as a faster moving planet passed by it in its orbit about the Sun.
7. Ptolemy—Earth; Copernicus—Sun. The Moon orbited Earth in both models.
8. Anywhere on Earth's equator.
9. Anywhere on Earth's equator. This occurs because at the equator the celestial poles make an angle of zero with the horizon. Since circumpolar stars are within the angle between a celestial pole and the horizon, there is no angle range for the circumpolar stars to exist in.
10. Since the zenith always makes a 90° angle to due south along the meridian, the celestial equator makes an angle of $90 - 80 = 10°$ to the zenith along the meridian. This means that the observer is at a latitude of 10° north.

Chapter Four

1. 0.469 au
2. Since an elliptical orbit is symmetrical about the semiminor axis, at the point of perihelion, the distance between the point and the empty focus is simply the aphelion distance, and by definition the distance between the point and the Sun is the perihelion distance. Adding Eqs. (4.2) and (4.3) yields $a(1 - e) + a(1 + e) = a - ae + a + ae = 2a$

3. This statement is not true because Kepler's second Law states that an area swept out by a line drawn from a planet to the Sun sweeps out equal areas in equal times. Both times are equal, but it can be seen in the figure that the distances from the planet to the empty focus are closer to the planet than the distances from the planet to the Sun during the time the second area was swept out. This means that the area swept out when Mercury moves from point C to point D would be smaller than A_1.

4. According to Kepler's third law, the orbital period and semimajor axis are not directly proportional to each other, but rather the orbital period squared is directly proportional to the semimajor axis cubed. Therefore, doubling the semimajor axis would not double the orbital period. Since $P^2 = a^3$, a body in an orbit with a semimajor axis of 2 au would have an orbital period of $P = 2^{\frac{3}{2}} = 2.83$ years.

5. According to Kepler's third law, the orbital period squared is directly proportional to the semimajor axis cubed or $P^2 = ka^3$. In Diagram 1 the semimajor axis is 2 au. In Diagram 2 the semimajor axis is 3 au. Taking the ratio yields:

$$\frac{P_2^2}{P_1^2} = \frac{a_2^3}{a_1^3}$$

$$\frac{P_2}{P_1} = \left(\frac{3}{2}\right)^{\frac{3}{2}} = 1.84$$

6. Using Eq. (4.7), M_p is replaced by m since now m is the body causing the tidal force. r is replaced with $\frac{R_p}{2}$. Dividing by the mass of the particle yields
$$a_{tidal} = \frac{2G_c m}{d^3} \frac{R_p}{2} = \frac{G_c m R_p}{d^3}$$

7. From left to right these would be (a) A,B,C,D, (b) A,B,C,D, and (c) Same for all because they all have the same semimajor axis of 4 au. From Kepler's third law if the semimajor axis is constant, then so is the orbital period.

8. These would be Parabolic $v_\infty = 0$

   Hyperbolic $v_\infty > 0$
   Elliptical v_∞ is undefined

9. These would be

Resonance	Location (au)	Order
Neptune 3:2	22.95	1
Saturn 5:9	14.1	4
Jupiter 1:3	10.8	2

10. First, find the location of the 1:2 mean motion resonance of Uranus using Eq. (4.45). Then subtract this location from the semimajor axis of Neptune's orbit. The result is $19.2\left(\frac{2}{1}\right)^{(2/3)} - 30.1 = 0.38$ au

11. Dividing 84.1 by 29.5 yields 2.85, which rounds up to 3 so the resonance would be 3:1.
12. These would be (a) A,B,C,D, (b) A,B,C,D, and (c) Same for all. The semimajor axis is 4 au.
13. 3.02
14. The 4:3 MMR of Uranus. This is found using trial and error.
15. About three cycles are completed in 9×10^5 years, so the period is 300,000 years which matches the period of the longitude of perihelion of Jupiter's orbit as found in Table 4.3. Thus, the test particle would be in the ν_5 resonance.
16. The resonances are (A) Kozai-Lidov, (B) ν_8, (C) ν_{16}, and (D) ν_{17}
17. The resonances are (A) 2:1, (B) 3:2, (C) 4:3, and (D) 3:1
18. The width is about 0.4 au. The maximum value of the eccentricity is 0.2.
19. 3×10^6 years

Chapter Five

1. The answers are

Region	Answer
Centaurs	C
Jupiter Trojans	B
Main Asteroid Belt	A
Trans-Neptunian objects	D
Edgeworth-Kuiper Belt	D
Origin of comets	D

2. These are:

Body	SSSB	Minor planet	Dwarf planet
Halley's Comet	x		
Pluto		x	x
An Aten	x	x	
Titan			

3. The answers are

Body	
95P Chiron	<200 years
3D Biela	No longer exists
1P Halley	<200 years
85D Boethin	No longer exists
C/1995 O1, Hale-Bopp	>200 years

4. Most Main Belt Asteroids are on the order of 1 km across and made of rocky material.

5. Though both types have semimajor axes greater than Earth's, Apollo asteroids cross Earth's orbit and Amor asteroids do not.

6. Comets have more ice than asteroids and usually have more eccentric orbits.

7. the coma

8. A comet is "active" when it outgasses due to sublimation of ices.

9. Turning directly from a solid into a gas. This is important because the activity of comets is due to sublimation of ices.

10. (A) water ice and (B) other ices such as carbon monoxide escape from below the surface due to the transformation of amorphous ice into crystalline ice.

11. Chiron has been found to be active at distances beyond 3 au. At such larger distances, water-ice is too cold to sublimate.

12. Because Centaurs have characteristics of both comets and asteroids. They can contain large amounts of ice like comets and can be active, but like asteroids have orbits with relatively lower eccentricities than that of traditional comets.

13. 10^6 years

14. the answers are

Designation	i (°)	e	a (au)	Heba major	Heba minor
2018 VG18	31.7	0.772	95.234	UT	III
2018 RR2	40.2	0.641	21.408	SE	II
2018 EZ1	29.7	0.393	16.26	SN	III
2017 YG5	23.7	0.88	62.233	ST	II
2017 WW14	6	0.747	22.747	JE	IV
2017 UX51	90.5	0.747	30.111	SE	I
2016 FX59	29.6	0.06	42.462	EK	

15. The answers are

Designation	a (au)	e	Comet class
Halley	17.94	0.967	I
21P Giacobini-Zinner	3.52	0.706	SP
Hyakutake	1165	0.9998	L
75P Kohoutek	3.4	0.537	SP
81P Wild 2	3.44	0.540	SP

16. The answers are

Designation	a (au)	e	i (°)	Levison comet class
Halley	17.94	0.967	162.2	Halley type
21P Giacobini-Zinner	3.52	0.706	31.8	Jupiter-family
Chariklo	15.8	0.172	23.4	Chiron type
Hyakutake	1165	0.9998	124.9	External
81P Wild 2	3.44	0.540	3.2	Jupiter-family

Chapter Six

1. $R_{td} \approx 2.67 \times 10^5$ km
2. 1.05 Earth radii
3. The largest change in orbital eccentricity of any ring particle orbit be ≥ 0.01.
4. The ring limit should increase with increasing ring orbital radius because according to Eq. (4.7), tidal forces due to the planet increase with ring orbital radius. Therefore, Body C will have the largest ring limit in when it is measured in kilometers. According to Fig. 6.4, the ring limit expressed in units of tidal disruption distances also increases with ring orbital radius, so again the answer is Body C.
5. 1.8 R_{td}; a parabolic orbit
6. The tidal disruption distance is 2.53×10^6 km. The Hill radius would be $25.7 R_{td}$ and the Roche limit $0.0215 R_{td}$.
7. The severity of each close encounter would be

v_∞ (km/s)	d_{min} (R_{td})	Severity
0	2.5	Low
3	1.5	Moderate
5	0.003	Extreme
9	30	Very low

8. $\delta = 0.199$ and $C = 2.06 R_{td}$. The regression coefficient is 0.975. The contour plateaus at high v_∞ and contours of Saturn do not display this behavior.
9. The slope is 0.254 $\frac{R_{td}}{\text{km/s}}$ and the y intercept is 1.74 R_{td}. The regression coefficient is 0.998, which is more accurate than that of the previous problem.

References

Agnor, Craig B., & Hamilton, Douglas P. 2006, Nature, 441, 192

Altenhoff, W. J., Menten, K. M., & Bertoldi, F. 2001, A&A, 366, L9

Altwegg, K., Balsiger, H., Bar-Nun, A., Berthelier, J. J., Bieler, A., Bochsler, P., Briois, C., Calmonte, U., Combi, M., De Keyser, J., Eberhardt, P., Fiethe, B., Fuselier, S., Gasc, S., Gombosi, T. I., Hansen, K. C., Hässig, M., Jäckel, A., Kopp, E., Korth, A., LeRoy, L., Mall, U., Marty, B., Mousis, O., Neefs, E., Owen, T., Rème, H., Rubin, M., Sémon, T., Tzou, C.-Y., Waite, H., & Wurz, P. 2015, Science, 347, article id. 1261952

Araujo, R. A. N., Sfair, R., & Winter, O. C. 2016, ApJ, 824, article id. 80

Araujo, R. A. N., Winter, O. C., & Sfair, R. 2018, MNRAS, 479, 4770

Aron, J. 2013, New Scientist, 216(2911), 6

Bailey, B. L., & Malhotra, R. 2009, Icarus, 203, 155

Batygin, K., & Morbidelli, A. 2017, AJ, 154, 229

Benz, W., Anic, A., Horner, J., & Whitby, J. A. 2008, Mercury, Space Sciences Series of ISSI, Volume 26. ISBN 978-0-387-77538-8. Springer Science+Business Media, BV, 2008, p. 7, 7

Bouchard, M. C., Howell, S. M., Chou, L., et al. 2018, Lunar and Planetary Science Conference, 2087

Braga-Ribas, F., Sicardy, B., Ortiz, J. L., et al. 2014, Nature, 508, 72

Brasser, R., Schwamb, M. E., Lykawka, P. S., & Gomes, R. S. 2012, MNRAS, 420, 3396

Brown, M. E. 2017, AAS/Division for Planetary Sciences Meeting Abstracts, 49, 405.06

Burbine, Thomas, H. 2017, Asteroids: Astronomical and Geological Bodies (Cambridge)

Burns, J. A., Showalter, M. R., Hamilton, D. P., et al. 1999, Science, 284, 1146

Bus, S. J., A'Hearn, M. F., Bowell, E., & Stern, S. A. 2001, Icarus, 150, 94

Bus, S. J., A'Hearn, M. F., Schleicher, D. G., & Bowell, E. 1991, Science, 251, 774

Bus, S. J., Bowell, E., & French, L. M. 1988, IAUC, 4684, 2

Bus, S. J., Bowell, E., Harris, A. W., & Hewitt, A. V. 1989, Icarus, 77, 223

Bus, S. J., Buie, M. W., Schleicher, D. G., et al. 1996, Icarus, 123, 478

Canup, R. M. 2005, Science, 307, 546

Capria, M. T., Coradini, A., De Sanctis, M. C., & Orosei, R. 2000, AJ, 119, 3112

Chambers, J. E. 1999, MNRAS, 304, 793

Cincotta, P. M., Giordano, C. M., & Simó, C. 2003, Physica D Nonlinear Phenomena, 182, 151

Cincotta, P. M., & Simó, C. 2000, A&A, 147, 205

Connors, M., Wiegert, P., & Veillet, C. 2011, Nature, 475, 481

de la Fuente Marcos, C., & de la Fuente Marcos, R. 2014, Ap&SS, 352, 409

de la Fuente Marcos, C. & de la Fuente Marcos, R. 2014, MNRAS, 439, 2970

De Luise, F., Dotto, E., Fornasier, S., et al. 2010, Icarus, 209, 586

Di Sisto, R. P., & Brunini, A. 2007, Icarus, 190, 224

Dones, L., Brasser, R., Kaib, N., & Rickman, H. 2015, Space Sci. Rev., 197, 191

Dones, L., Levison, H. F., & Duncan, M. 1996, Completing the Inventory of the Solar System, 107, 233

Duffard, R., Lazzaro, D., Pinto, S., et al. 2002, Icarus, 160, 44

Duncan, M. J., Levison, H. F., & Budd, S. M. 1995, AJ, 110, 3073

Duncan, M., Levison, H., & Dones, L. 2004, Comets II, 193

Duncan, M., Quinn, T., & Tremaine, S. 1987, AJ, 94, 1330

Duncan, M., Quinn, T., & Tremaine, S. 1988, ApJL, 328, L69

Elliot, J. L., Olkin, C. B., Dunham, E. W., et al. 1995, Nature, 373, 46

Ellis, K. M., & Murray, C. D. 2000, Icarus, 147, 129

El Moutamid, Maryame, Kral, Quentin, Sicardy, Bruno, Charnoz, Sebastien, Roques, Françoise, Nicholson, Philip D., & Burns, Joseph A. 2014, American Astronomical Society, DDA meeting #45, #402.05

Emel'yanenko, V. V., Asher, D. J., & Bailey, M. E. 2005, MNRAS, 361, 1345

Emel'yanenko, V. V., Asher, D. J., & Bailey, M. E. 2013, Earth Moon and Planets, 110, 105

Festou, M. C., Keller, H. U., & Weaver, H. A. 2004, Comets II

Fornasier, S., Lazzaro, D., Alvarez-Candal, A., et al. 2014, A&A, 568, L11

Fornasier, S., Lellouch, E., Müller, T., et al. 2013, A&A, 555, A15

Fouchard, M., Rickman, H., Froeschlé, C., et al. 2014, Icarus, 231, 110

French, R. G., McGhee, C. A., Dones, L., & Lissauer, J. J. 2003, Icarus, 162, 143

French, R. G., Nicholson, P. D., McGhee-French, C. A., et al. 2016, Icarus, 274, 131

Froeschle, C., & Morbidelli, A. 1994, Asteroids, Comets, Meteors 1993, 160, 189

Froeschle, C., & Scholl, H. 1986, A&A, 166, 326

Froeschle, C., & Scholl, H. 1989, Celestial Mechanics and Dynamical Astronomy, 46, 231

Gallardo, T. 2006, Icarus, 184, 29

Giordano, C. M., & Cincotta, P. M. 2004, A&A, 423, 745

Gleik, J. 1987, Chaos: Making a New Science (Penguin Publishing Group)

Gomes, R. S. 2003, Icarus, 161, 404.

Gomes, R., Levison, H. F., Tsiganis, K., & Morbidelli, A. 2005, Nature, 435, 466

Goździewski, K., Bois, E., Maciejewski, A. J., & Kiseleva-Eggleton, L. 2001, A&A, 378, 569

Groussin, O., Lamy, P., & Jorda, L. 2004, A&A, 413, 1163

Gurfil, Pini, Seidelmann, Kenneth, P. 2016, Celestial Mechanics and Astrodynamics: Theory and Practice (Springer)

Hahn, G., & Bailey, M. E. 1990, NATURE, 348, 132

Hairer, Ernst, Nørsett, Syvert P., & Wanner, Gerhard 1993, Solving Ordinary Differential Equations I, Vol. 8, (Springer)

Hartmann, W. K., Tholen, D. J., Cruikshank, D. P., Brown, R. H., & Morrison, D. 1988, BAAS, 20, 836

Hartmann, W. K., Tholen, D. J., Meech, K. J., & Cruikshank, D. P. 1990, Icarus, 83, 1

Hedman, M. M., Burns, J. A., Tiscareno, M. S., et al. 2007, Science, 317, 653

Heisler, J., & Tremaine, S. 1986, Icarus, 65, 13

Hill, G. W. 1878, Am. J. Math, 1, 5

Hills, J. G. 1981, AJ, 86, 1730

Hinse, T. C., Christou, A. A., Alvarellos, J. L. A., & Goździewski, K. 2010, MNRAS, 404, 837

Holman, M. J., & Wisdom, J. 1993, AJ, 105, 1987

Horner, J., Evans, N. W., & Bailey, M. E. 2004, MNRAS, 354, 798

Horner, J., Evans, N. W., & Bailey, M. E. 2004, MNRAS, 355, 321

Horner, J., Evans, N. W., Bailey, M. E., & Asher, D. J. 2003, MNRAS, 343, 1057

Horner, J., & Jones, B. W. 2010, Astronomy and Geophysics, 51, 6.16

Horner, J., & Lykawka, P. S. 2010, MNRAS, 402, 13

Horner, J. & Wyn Evans, N. 2006, MNRAS, 367, L20

Hyodo, Ryuki, Charnoz, Sébastien, Genda, Hidenori, & Ohtsuki, Keiji 2016, ApJL, 828, L8

Jang-Condell, H. 2017, APJ, 835, 12

Jewitt, David 2009, AJ, 137, 4296

Jewitt, D., & Luu, J. 1996, Completing the Inventory of the Solar System, 107, 255

Kirkwood D., 1867, Meteoric Astronomy. J. B. Lippincott, Philadelphia

Kortenkamp, S. J., Malhotra, R., & Michtchenko, T. 2004, Icarus, 167, 347

Kowal, C. T., Liller, W., & Marsden, B. G. 1979, Dynamics of the Solar System, 81, 245

Kozai, Y. 1962, AJ, 67, 591

Lacerda, P., Jewitt, D., & Peixinho, N. 2008, Asteroids, Comets, Meteors 2008, 1405, 8007

Laskar, J., & Boué, G. 2010, A&A, 522, A60

Lazzaro, D., Florczak, M. A., Angeli, C. A., et al. 1997, Planet. Space Sci., 45, 1607

Lazzaro, D., Florczak, M. A., Betzler, A., et al. 1996, Planet. Space Sci., 44, 1547

Lellouch, E., Kiss, C., Santos-Sanz, P., et al. 2010, A&A, 518, L147

Levison, H. F. 1996, Completing the Inventory of the Solar System, 107, 173

Levison, H. F., & Duncan, M. J. 1993, ApJL, 406, L35

Levison, H. F., & Duncan, M. J. 1994, Icarus, 108, 18

Levison, H. F., & Duncan, M. J. 1997, Icarus, 127, 13

Levison, H. F., Duncan, M. J., Dones, L., & Gladman, B. J. 2006, Icarus, 184, 619

Lewis, M. C., & Sickafoose, A. A. 2017, AAS/Division for Planetary Sciences Meeting Abstracts #49, 220.01

Lockwood, A. C., Brown, M. E., & Stansberry, J. 2014, Earth Moon and Planets, 111, 127

Luu, J. X. 1993, PASP, 105, 946

Luu, J. X., & Jewitt, D. C. 1990, AJ, 100, 913

Lykawka, P. S., Horner, J., Jones, B. W., & Mukai, T. 2011, MNRAS, 412, 537

Lykawka, P. S., & Mukai, T. 2007, Icarus, 189, 213

Malhotra, R. 1994, Physica D Nonlinear Phenomena, 77, 289

Malhotra, R. 1995, AJ, 110, 420

Malhotra, R., & Dermott, S. F. 1990, Icarus, 85, 444

Marzari, F. & Scholl, H. 2000, Icarus, 146, 232

Marzari, F., & Scholl, H. 2002, Icarus, 159, 328

Marzari, F., Tricarico, P. & Scholl, H. 2002, APJ, 579, 905

Masaki, Y., Kitasato, Tsukuba, & Kinoshita, H. 2003, A&A, 403, 769

Meech, K. J., & Belton, M. J. S. 1990, AJ, 100, 1323

Meech, K. J., Buie, M. W., Samarasinha, N. H., Mueller, B. E. A., & Belton, M. J. S. 1997, AJ, 113, 844

Meech, K. J., & Svoren, J. 2004, Comets II, 317

Melita, M. D., Duffard, R., Ortiz, J. L., & Campo-Bagatin, A. 2017, A&A, 602, A27

Michikoshi, S., & Kokubo, E. 2017, ApJL, 837, L13

Moons, M., & Morbidelli, A. 1995, Icarus, 114, 33

Moons, M., Morbidelli, A., & Migliorini, F. 1998, Icarus, 135, 458

Morbidelli, A. 2002, Modern celestial mechanics: aspects of solar system dynamics

Murray, C.D., & Dermott, S. F. 1999, Solar system dynamics (Cambridge University Press)

Murray-Clay, R. A., & Schlichting, H. E. 2011, ApJ, 730, 132

Napier, B., Asher, D., Bailey, M., & Steel, D. 2015, Astronomy and Geophysics, 56, 6.24

Nesvorný, D., & Dones, L. 2002, Icarus, 160, 271

Norton, O. R., & Chitwood, L. A. 2008, Field Guide to Meteors and Meteorites: Patrick Moore's Practical Astronomy Series. (London: Springer-Verlag)

Oort, J. H. 1950, Bulletin of the Astronomical Institutes of the Netherlands, 11, 91

Ortiz, J. L., Duffard, R., Pinilla-Alonso, N., et al. 2015, Ap&SS, 576, A18

Ortiz, J. L., Santos-Sanz, P., Sicardy, B., et al. 2017, Nature, 550, 219

Pan, Margaret, & Wu, Yanqin 2016, AJ, 821, article id. 18

Perna, D., Dotto, E., Barucci, M. A., et al. 2009, A&A, 508, 451

Petit, J.-M., Morbidelli, A., & Chambers, J. 2001, Icarus, 153, 338

Philpott, C. M., Hamilton, D. P., & Agnor, C. B. 2010, Icarus, 208, 824

Planetary Laboratory, Purple Mountain Observatory, & Stellar Division of Peking Observatory 1978, Chinese Astronomy, 2, 341

Porco, C. C., Baker, E., Barbara, J., et al. 2005, Science, 307, 1226

Potashko, Oleksandr, & Viso, Michel 2014, 40th COSPAR Scientific Assembly, 40, B0.6-5-14

Prialnik, D., Brosch, N., & Ianovici, D. 1995, MNRAS, 276, 1148

Quinn, T., Tremaine, S., & Duncan, M. 1990, ApJ, 355, 667

Rabinowitz, D. L., Barkume, K., Brown, M. E., et al. 2006, ApJ, 639, 1238

Rein, H., & Spiegel, D. S. 2015, MNRAS, 446, 1424

Rein, H., & Tamayo, D. 2015, MNRAS, 452, 376

Rimlinger, T., Hamilton, D. P., & Hahn, J. M. 2017, AAS/Division for Planetary Sciences Meeting
      Abstracts, 49, 501.03

Roig, F., Nesvorný, D., & Ferraz-Mello, S. 2002, MNRAS, 335, 417

Ryden, Barbara, S. 2016, Dynamics (The Ohio State University)

Scholl, H., Marzari, F., & Tricarico, P. 2005, Icarus, 175, 397

Schubart, J. 2007, Icarus, 188, 189

Sekanina, Z. 1996, A&A, 314, 315

Sfair, R., Araujo, R., & Winter, O. C. 2018, AAS/Division for Planetary Sciences Meeting
      Abstracts, 315.05

Sfair, R., private communication

Sheppard, S. S., & Trujillo, C. A. 2006, Science, 313, 511

Shevchenko, Ivan, I. 2017, The Lidov-Kozai Effect - Applications in Exoplanet Research and
      Dynamical Astronomy (Springer International Publishing)

Sicardy, B., Leiva, R., Renner, S., et al. 2019, Nature Astronomy, 3, 146

Sie, Z.-F., Lin, H.-W., & Ip, W.-H. 2015, AAS/Division for Planetary Sciences Meeting Abstracts
      #47, 47, 211.10

Silva, A. M., & Cellone, S. A. 2001, Planet. Space Sci., 49, 1325

Smirnov, E. A., & Shevchenko, I. I. 2013, Icarus, 222, 220

Spahn, F., Petit, J.-M., & Bendjoya, P. 1993, Celestial Mechanics and Dynamical Astronomy,
      57, 391

Stern, S. A. 1989, PASP, 101, 126

Stern, S. A. 1995, S&T, 89, 32

Sykes, M. V., & Walker, R. G. 1991, Science, 251, 777

Tabeshian, M., & Wiegert, P. A. 2016, APJ, 818, 159

Thirouin, A., Sheppard, S. S., Noll, K. S., et al. 2016, AJ, 151, 148

Tiscareno, M. S., & Malhotra, R. 2003, AJ, 126, 3122

Tiscareno, M. S., & Malhotra, R. 2009, AJ, 138, 827

Volk, K., & Malhotra, R. 2008, ApJ, 687, 714–725

Weaver, H. A., & Lamy, P. L. 1997, Earth Moon and Planets, 79, 17

Whipple, A. L. 1995, Icarus, 115, 347

Williams, J.G. 1969, Ph.D. dissertation,University of California, Los Angeles

Williams, J. G., & Benson, G. S. 1971, AJ, 76, 167

Wisdom, J. 1982, AJ, 87, 577

Wisdom, J. 1983, Icarus, 56, 51

Womack, M., Sarid, G., & Wierzchos, K. 2017, Publications of the Astronomical Society of the
      Pacific, 129, 31001

Womack, M., & Stern, S. A. 1995, BAAS, 27, 33.07

Wood, J., Horner, J., Hinse, T. C., & Marsden, S. C. 2017, AJ, 153, 245

Wood, J., Horner, J., Hinse, T. C., & Marsden, S. C. 2018, AJ, 155, 2

Wood, J., Horner, J., Hinse, T. C., et al. 2018, MNRAS, 480, 4183

Printed in the United States
By Bookmasters